D0779210

Of Two Minds

Studies in Literature and Science
published in association with the Society for Literature and Science

Transgressive Readings: The Texts of Franz Kafka and Max Planck
by Valerie D. Greenberg

A Blessed Rage for Order: Deconstruction, Evolution, and Chaos
by Alexander J. Argyros

Of Two Minds: Hypertext Pedagogy and Poetics by Michael Joyce

The Artificial Paradise: Science Fiction and American Reality
by Sharona Ben-Tov

Conversations on Science, Culture, and Time
by Michel Serres with Bruno Latour

Genesis by Michel Serres

The Natural Contract by Michel Serres

Dora Marsden and Early Modernism: Gender, Individualism, Science
by Bruce Clarke

The Meaning of Consciousness by Andrew Lohrey

The Troubadour of Knowledge by Michel Serres

Simplicity and Complexity: Pondering Literature, Science, and Painting
by Floyd Merrell

Othermindedness: The Emergence of Network Culture
by Michael Joyce

Embodying Technesis: *Technology Beyond Writing* by Mark Hansen

Dripping Dry: Literature, Politics, and Water in the Desert Southwest
by David N. Cassuto

Networking: Communicating with Bodies and Machines in the Nineteenth Century by Laura Otis

The Knowable and the Unknowable: Modern Science, Nonclassical Thought, and the "Two Cultures" by Arkady Plotnitsky

Rethinking Reality: Lucretius and the Textualization of Nature
by Duncan F. Kennedy

Of Two Minds

Hypertext Pedagogy and Poetics

Michael Joyce

Ann Arbor

The University
of Michigan Press

First paperback edition 1996
Copyright © by the University of Michigan 1995
All rights reserved
Published in the United States of America
by the University of Michigan Press
Manufactured in the United States of America
♾ Printed on acid-free paper

2002 5 4 3

*A CIP catalog record for this book is
available from the British Library.*

No part of this publication may be reproduced,
stored in a retrieval system, or transmitted
in any form or by any means electronic, mechanical,
or otherwise, without the written permission
of the publisher.

Joyce, Michael, 1945–
Of two minds : hypertext pedagogy and poetics / Michael Joyce
p. cm. – (Studies in literature and science)
Includes bibliographical references.
ISBN 0-472-09578-1 (alk. paper)
1. Literature and technology. 2. Hypertext systems.
3. Hypermedia systems.
4. Literature—Study and teaching. 5. Poetics. I. Title II. Series.
PN56.T37J69 1995
800—dc20 94-23738
CIP

ISBN 0-472-06578-5 (pbk. : alk. paper)

Acknowledgments

I believe the thanks show through in the polylogue of this text. If not, I pray that those whom I have to thank look for themselves in the words they love best and accept my apologies with my love. Beyond this, though, there are a few whom I must mention because they are my other selves:

For years I have been unable to tell where Jay Bolter's thought leaves off and mine begins, and my life has grown rich on this account.

Martha Petry is a brilliant teacher of writers and readers, and so she often knows the inner rhythms of my voice better than I do. I have gained whenever I have the grace to hear her or to reorganize paragraphs, or my life, according to her eye. She wrote indelibly upon my life and heart and for years wore my name.

Our boys, Eamon Paul Joyce and Jeremiah David Joyce, have written themselves upon all my texts and suffered the distance it sometimes took to be able to see.

My brothers and sisters—Kathy, Susan, Tom, Mary Ethel, Rosemary, Brian and Tim—are a tribe. Always among us is the memory of my mother, Joanne Poth Joyce, a proud woman, and my father, Thomas Robert Joyce, a caring man. We believe they live in our words.

My colleagues and friends at Vassar College—especially Barbara Page, Paul Kane, Beverly Haviland, and Dean Nancy Schrom Dye—gave me a home when it was most needed and I was "Heimatlos" as my friend the poet, John Enright, wrote: "Heimatlos / we are *vagi* / our marriage is such."

Finally, there is salt for the barefoot girl in her forest castle, chocolate for the mad girl in her books, spices for the woman by the frozen lake.

Contents

Introduction: The Comfort of Knowing We Are Not Lost

A story is told about two elderly sisters who made a pact that the first to die would, as her death approached, begin talking to the other sister at her bedside. The dying sister was to tell everything she heard and saw, attempting to talk without ceasing up to and, if at all possible, *past* the point of death. The sisters did not dare to hope that they could penetrate the dark edge of existence. Rather, they wished to offer the comfort of knowing they were not lost to each other at the end of time. Their pact was thwarted, however, when the first sister choked to death on a clump of ice accidentally aspirated as she drank ice water. Wide eyed with dismay at the unexpectedness of it all, facing both her sister and her death, and of course unable to speak, she was nonetheless bemused at suffocating on something in the process of disappearing.

This comic parable, which occurs in a twenty-some-year-old (and a twenty-some-year-old's) unpublished novel of mine, seems now to bear upon the situation we face in what Jay Bolter calls "the late age of print." We too are late at the end of something and unable to speak. It is too much to hope for an orderly enough end to accomplish our hopes of transcendence. Yet what chokes and thwarts us is so transitory that to hope for justice seems foolish. In a young man's novel such a parable serves to illuminate the absurdity of a young man's world; in the adult's collection of essays it is offered as an expression of belief in the comfort of knowing we are not lost to one another. This is called a change of perspective.

Much of what follows is about changed perspectives, or what my mentor, Sherman Paul, terms perspectivalization, a primary postmodern virtue. In 1982 I bought a microcomputer, a decision that proceeded from my interests and identity as a writer-teacher. Immediately, as in the most naive visions of pulp science fiction—and despite my claims that this perspectivalizing machine was only a tool, a way of knowing—the computer began to change me. Possessed of two minds, my own and its augmented silicon, I began the slow process of learning to see this not so newly doubled self as one; to see, as Carolyn Guyer has put it, that "dualities must be in tensional opposition to each other in order for the central paradox of existence to work."

Slowly I came to recognize myself veering toward becoming something of a cyborg, in Donna Haraway's sense, a creature whose life is no longer bounded by the horizon line that slashes between too comfortable dualisms such as "self/other, mind/body, culture/nature, male/female, civilized/primitive, reality/appearance, whole/part, agent/resource, maker/made. . . ." Previously stable horizons across my psychic landscape gave way to dizzying patterns of successive contours, each of which was most assuredly real, each of which did not last. Because previous landmarks would not stand still, I have had to learn to measure my progress differently, becoming in the process an example of my mother's distinctive version of the cyborg: "Those dizzy bastards who claim that just because they're spinning the world has changed."

Spinning, the world does change. "Bound in a spiral dance," in Haraway's phrase, we are left dizzied and wondering how to use what we know in order to make our way. So shaken, we suddenly are able "to see," as Don Byrd characterizes the vision of the poet Charles Olson, "that the content of the world has uses inside the skin." A pirouetting dancer "spots" a point in space to avoid dizziness; a pilot uses a horizon indicator to turn a barrel roll. Yet it is the body that not only knows the change but literally makes the way.

This is to say that change is the only true destination, the only reliable occupation in a world that, as Haraway says, "make[s] problematic the statuses of man or women, human, artifact, member of race, individual identity, or body." Olson uses the word *proprioception* to account for the uses of the content of the world inside the skin. Proprioception is the body's knowledge of its own depth and location, its internalized perspective of "how to use oneself and on what," in Olson's phrase. Yet graphical computer interfaces, hypertexts, virtual realities, and other instances of what Haraway calls the "couplings between organism and machine conceived as coded devices" serve to externalize the internal. Such varieties of technological experience insist upon the permeability of the "self" and the "what" and the uses made of each. Cyborg consciousness invites us to turn proprioception outward beyond Haraway's "crucial boundary breakdown" of organism-machine to a place where we take "pleasure in the confusion of boundaries and responsibility in their construction."

The essays collected here talk sometimes confusingly, although, I hope, also pleasurably and permeably, about teaching and writing—hypertext pedagogy and poetics—from the vantage of internalized perspectives about each. Much of this collection takes the form of what, aware of the ambiguity, I call theoretical narratives, i.e., essays that are both theoretically in narrative form and also narratives of theory. These narratives are made up in the way that a bed is, a day-to-day process of billowing, shaking, refitting. Many of them were talks before they became essays, not so unusual an occurrence for such a collection. That they stay "talks" here should also not seem unusual in an age of virtual realities and telepresences.

What is unusual (but increasingly characteristic of the late age of print) is that, before, during, and after they were talks or essays, these narratives were often e-mail [electronic mail] messages, hypertext "nodes," and other kinds of electronic text that, as will be seen, moved nomadically and iteratively from one talk to another, one draft to another, one occasion or perspective to another. The nomadic movement of ideas is made effortless by the electronic medium that makes it easy to cross borders (or erase them) with the swipe of a mouse, carrying as much of the world as you will on the etched arrow of light that makes up a cursor. At each crossing a world of possibility can be spewed out in whole or in kernel, like the cosmogonic dragon's teeth of myth. Each iteration "breathes life into a narrative of possibilities," as Jane Yellowlees Douglas says of hypertext fiction, so that, in the "third or fourth encounter with the same place, the immediate encounter remains the same as the first, [but] what changes is [our] understanding." The text becomes a present tense palimpsest where what shines through are not past versions but potential, alternate views.

I have not attempted to yield design to these alternate views and to fashion the organization of this text into a hypertextual analogue in the manner of texts like *S/Z* or *Glas* or the more recent *Telephone Book.* Partly this is because I am by now used to readers who insist that my prose is so polyvocal that I speak or write hypertext, even (or especially) each time that I am certain I have finally, like a line upon water, written a clear version. Also, there is a certain joy to the sedimentary evidence of the equivalence between seeing and knowing which years of text in all its soarings and its failings nonetheless at some times the book best represents. Last, there is a pleasure in writing a book about their end(s), which, like Haraway's cyborg, "skips the original unity" and is "resolutely committed to partiality, irony, intimacy, and perversity."

It seems perversely right that a book titled *Of Two Minds* should have three parts, though I have an anonymous publisher's reviewer to thank for leading me to see the obvious virtues of pairing introductory essays within their own section contextualizing the major concerns of this collection. Setting two tonal essays in harmonic juxtaposition makes it easy, if not to skip, at least to subvert the original unity duality represents. Dual channels give way to something more like the permeable flow of meaning between sometimes veering, sometimes nearing, banks of a single river. "Within the analytic tradition that parses complex flow as combinations of separate factors," as N. Katherine Hayles observes, "it is difficult to think complexity. . . practitioners forget that in reality there is always only the interactive environment as a whole."

Yet, since a certain parsing into separate factors is inevitable and even useful in plumbing a book of this sort, it seemed best to follow the same anonymous reviewer's suggestions for occasionally reminding the reader of the interactive environment as a whole by throwing the river into contrary eddies. Thus, the two major sections here are interspersed with "interstitial" documents, a name borrowed from my own term (in "A Feel for Prose," chap. 14) for the capillary flow of interaction. To say that these interstitials are documents probably gives them more historical weight than they should have (at least one such section is, however, an attempt at herstory). Dispatches, apologia, meditations, are all closer to the mark; fretwork summons the right kind of troubled ambiguity. More occasional than the longer occasional pieces collected here, the interstitials are in fact public statements requested for various settings, a few inward and about my work, a few outward and about others; some poetical, some didactic, some antic. Pedagogic interstitials are plopped down with poetic essays, and vice versa, with one set of both placed on the cusp between major sections of the book. They are set here like stones dragged from opposite streams, meant to weave the water, turning it more complex in its course.

The essays in this book are meant to weave two minds, hypertext pedagogy and poetics, into proprioceptive soundings, the momentary wholeness of changing understandings. Like the sisters of the young man's story, both pedagogy and poetics—teaching and writing—have come almost simultaneously to the end of very long strands of their development. As George Landow suggests, hypertext, with "[i]ts emphasis upon the active, empowered reader, which fundamentally calls into question general assumptions about reading, writing, and texts, similarly calls into question our assumptions about the literary education and its institutions that so depend upon these texts."

I have had the good fortune to work and write (separated in space by geographic distance though linked in time by networks, both computer and telephone) with a group of teaching writers and thinkers, the most of whom are (iteratively) cited throughout these essays. Together we have had the necessary pleasure of successively—not to say incessantly—questioning our assumptions. Marginalized, uncertain, ironic, paradoxical, playful, perverse, and lost, we have been fortunate in our opportunity (to adopt the postmodern typographical graphi-lect) to consider hypertext pedagogy and/*as* poetics in the process of re(de)fining each. Hypertext pedagogy and poetics alike form what Gilles Deleuze and Félix Guattari call "complex differences; the de facto mixes, and the passages from one to the other; the principles of the mixture, which are not at all symmetrical, sometimes causing a passage from the smooth to the striated, sometimes from the striated to the smooth."

Thus, these essays and/or narratives are less a collection than a concoction; shaped largely in what Carolyn Guyer calls the "buzz daze" of space-time and committed to what Jane Yellowlees Douglas calls "rejecting the objective paradigm of reality as the great 'either/or' and embracing, instead, the 'and/and/and.'" As a perspectivalizing machine (variously embodied as synchronous computer conferencing, collaborative networking, database searching, desktop publishing and presentation, hypertext, hypermedia, simulation and virtual reality), the computer not only extends the text to encompass the world but also likewise extends the fundamental questions about education and institutions that Landow raises. Our genuine culture, which is to say our experience of living in a place over time, is increasingly enacted not just in the manipulation of symbolic information but also in our increasing willingness to see our own existence as both constituted by and constituting symbolic information.

Part 1: Of Two Minds: Hypertext Contexts

The essay introducing hypertext pedagogy in "Hypertext and Hypermedia" was a commissioned entry for the *Encyclopedia on English Studies and Language Arts,* "Hypertext and Hypermedia." Not surprisingly, it is perhaps the most cogent, even traditional, essay here. As such it may serve as a litmus test for the wary reader, although the truth is, as the persistent reader will discover, that the stylistic calm of the encyclopedia gives strategic cover for some rather radical claims for hypertext, beginning with the first sentence, "Hypertext is, before anything else, a visual form." In any case this brief chapter introduces the issues, definitions, history, and contexts for hypertext.

"What I Really Wanted to Do I Thought," the essay introducing hypertext poetics, was gathered for Jon Lanestedt—the brilliant, young Norwegian hypertext theorist, programmer, and author—as background for an article he was preparing regarding Jay Bolter's and my hypertext program Storyspace. In some sense it transcribes a set of formulaic anecdotes I was used to telling in talks and readings, interviews and classes, but that I had never quite pulled together into any rigid sequence. It means to describe, as best I can, the befuddling process of coming to terms with the fact that one is supposed to have had a vision when it only seemed one did what one thought (which is three ones too many and so doesn't add up). This is to say, it attempts to answer the frequently asked, and characteristically American, question "How'd you ever think up something like that, anyway?"

Part 2: Siren Shapes: Hypertext Pedagogy

The essays regarding hypertext pedagogy begin in "Siren Shapes: Exploratory and Constructive Hypertext" with what is happily among the most frequently cited essays regarding hypertext pedagogy (so frequently cited that computer and composition doyenne—and self-styled "feminist cyborg guerilla"—Cindy Selfe once accused it, one hopes humorously, of being "canonical"). Unhappily, it appeared in a journal that is now defunct and so generally unavailable, at least in print. The definitions of exploratory and constructive hypertext that first emerge there make iterative appearances in later essays as, like in others, I struggle to understand their implications. This essay also reports some of the earliest work done with hypertext and developmental writing.

"Interstitial: Networks of Woven Water" and "The Dangers of Transparency." The first of these two statements was solicited as a contribution to a multiply authored manifesto culminating an electronic salon (conducted over the TechnoCulture [TNC] computer network mail list) preceding a seminar on cyberculture, coordinated by Ann Balsamo and Stuart Moulthrop at Georgia Tech. It means to recognize the formative importance of a female genealogy, in Luce Irigaray's Foucauldian phrase, for hypertext poetics. It is a weave, a blur, a ripple, a stream of successive, mutable fields, and, like teaching itself, a practical work.

The second statement was a contribution to what was meant to be a face-to-face (FTF, in "net" parlance) version of an electronic salon, a Conference on College Composition and Communication (CCCC) thinktank regarding computer intensive instruction (San Diego, 1993). It talks about when "the mirror metaphor fails and gives way to one of intervening, aqueous lenses whose translucence shifts from moment-to-moment between glass and mirror."

Also out of water comes by far the most curious, even awkward, essay here, "What the Fish Lady Saw: Patterns Out of Disjointedness in Two Hypertext Writing Communities." This essay is part of a continuing effort to address the "two cultures" problem in hypertext studies through a process of collision rather than conversation. Like its companion pieces (coauthored with Mark Bernstein and others for the Proceedings of the Association of Computing Machinery [ACM] but not collected here), it may represent the as yet unhappy marriage of bad science with bad literary-pedagogical theory. Like many bad marriages, however, its intent is earnest, even heroic: to find a way for the disparate concerns of scientist and humanist to meld together without sacrificing their particularity. The essay was prepared for a

CCCC workshop on research into computers and composition. Writers in two community college classes—a first-term composition class and a creative writing class—were invited to consider their experiences in networked, collaborative, hypertext writing environments by participating in synchronous conferences with other members of their learning community. Transcripts of the two computer conferences were organized into a Storyspace hypertext web then analyzed both through a narrative ethnography and with graphical representations derived from computer science, in order to characterize the instructor's view (i.e., mine) of writing communities and the issues that engaged them.

"Interstitial: Everyone's Story Seems to Go on Without Us" includes two short texts about hypertext fictions. The first was published as the introduction to Stuart Moulthrop's epic hypertext fiction, *Victory Garden*, and looks toward the work being done by Moulthrop and others establishing a hypertext literature and culture. The second, an introduction to Carolyn Guyer's undulant hyperfiction, *Quibbling*, remains unpublished. For whatever reason (perhaps on a theory about the epic and the undulant) Eastgate Systems issued her text in a jewel box, unadorned by frontmatter. Both these statements, in their concerns with multiple stories and what in another essay I call ongoingness, seem to fit nicely between the attempts to measure and represent ongoingness in the previous essay and the ongoingness of the multiple stories in the theoretical narratives of following chapters.

The concluding series of three essays in this section, "'So Much Time, so Little to Do': Empowering Silence and the Electric Book"; "A Memphite Topography"; and "New Teaching: Toward a Pedagogy for a New Cosmology," are examples of the bed-making theoretical narratives and iterative movements of nomadic ideas discussed earlier. The first two were keynote addresses, and the third ought to have been one instead of being published in a journal, at least according to the dissenting referee for *Computers and Composition*, the journal that did publish it as a lead essay. Consciously polemical, narrative, and meditative in style, these essays traced for me a movement from reciprocal silence to a city of text, from Willie Wonka to Wim Wenders, from marginal to multiple.

The three essays are nomadic in the sense of Deleuze and Guattari, in which "every point is a relay and exists only as a relay... [in which] the in-between has taken on all the consistency and enjoys both an autonomy and a direction of its own... [and the] life of the nomad is the intermezzo." The essays and their ideas remain in movement for me. With each iteration of them I believed I had come closer to, even achieved, a clear telling, a true story, a direct statement, and each time to my great dismay hearers and readers greeted (nay, even heralded) me with huzzahs for my hypertextual complexities. In lieu of the music of the spheres, I came to become fond of the poet Charles Bernstein's notion of "the music of contrasting characterizations." Although I have trimmed those cadenzas that are shared whole scale among the three pieces, enough of their pure repetition, iterative symmetry, and intermezzo remains either to annoy the reader or to make the movement clear depending on her ear for sarabandes and other spiral dances.

"Interstitial: Silicon Valley Maoists and Ohio Zen." Two lists of sixfold ways offer rules for the unruly and practical talk about impractical matters. The first, written in the late-twentieth-century poetic form called the "executive summary," offers "infotainment" merchants ways to talk to teachers and here serves to capitulate themes of the first half. The other, written in the late-twentieth-century dramatic form called "panelspeak," was meant to afford me a way to talk as if I were eloquent on the fly and off the cuff and serves to announce themes (and sometimes latterly appearing texts) of the second half. Both lists are more whispers than shouts, although here on the cusp between pedagogy and poetics they are meant to echo hither and thithering to and fro between converging shores.

Part 3: Contours: Hypertext Poetics

The essay that begins the second movement of this text, "Selfish Interaction: Subversive Texts and the Multiple Novel," was an actual first pass. I suggested above that any electronic text is a present-tense palimpsest; this one is made up of sections snatched from, among other sources, my first letters to Jay Bolter, our first grant request for Storyspace, and journal entries. It includes my earliest formulation of ideas about interactive fictions, intentionality, and author and reader relationships that are only finding their present-tense expression in recent speculations on hypertext contours. Fittingly for a palimpsest, it was promised for a collection of essays in 1986 but did not appear in print until 1991, when it had the curious fate of being reprinted in the McGraw-Hill *Hypertext Hypermedia Handbook* before it was printed in the Elsevier collection (which, in the way of academic texts, may now not appear at all). I still find myself challenged by its skepticism and puzzled by its claims, which often seem at variance with my more utopian mind of late.

"Interstitial: Dead White Men Also Compute" is perhaps an instance of my more utopian mind, though it too comes from rather early on, and all that is late in it are the eponymous white men of its title. It was initially prepared as a so-called white paper for an intrainstitutional, interdisciplinary grant writing team. "As a first step toward a hypertext for the humanities," it suggests that "we might imagine something very much like hyperfiction in conception or at least in its branching, convinced that the mix of seamless, default branches, yields, and patterned browsing which characterizes them is something like the right model for interactive texts in any domain, and especially for the kinds of choice-driven modules we are discussing." Its companion piece, "Network Culture (1994)," suggests that these issues persist into the Web and beyond.

"The Geography of the Word: The Textfile as Landscape" repays intellectual debts as it attempts to map ideational ground. Although it might perhaps be collected among the essays on hypertext pedagogy on account of its attention to the effect of technology upon the writing process, its concerns with artificial intelligence (AI) and the academic discipline and profession of geography each are grounded in the genealogy of a writer's consciousness. The linkage of the work of Roger Schank's group (with which I spent time as a visiting fellow at the Yale AI Lab) with the thought of the geographer Carl O. Sauer and the poet Charles Olson enacts the overt repayment. The real link, and the real debt, however, is to Sherman Paul, who still teaches me what it is to see.

"The Ends of Print Culture" appeared in the electronic journal *Postmodern Culture* (*PMC*) shrouded in a much longer title, which sandwiched its intention between cautionary pretexts and postscripts: "Notes toward an Unwritten Non-Linear Electronic Text, 'The Ends of Print Culture' (a work in progress)." The sandwiching took place in response to (again) the vehemence of referees who found it more polemical than philosophical (but who must have been mystified when

Robert Coover's even more polemically titled "The End of Books" appeared as the front-page essay of the *New York Times Book Review*). When editors Eyal Amiran and John Unsworth asked if I would mind publishing my essay as part of their proposed series of works in progress (which, to the best of my knowledge, includes only this essay, and which also, most likely, kept it out of the Oxford University Press collection of *PMC* essays), I was pleased, since, as will become evident, it is very much the fourth panel for the pedagogical triptych of polemical keynotes collected in the first section. As such, it gathers strands (and whole sections of nomadic text) from these three essays and uses them hypertextually to weave the helix of a new argument. Thus, I have neither edited its rather substantial borrowings, other than to italicize them (so a reader who tires of wholesale repetitions can pass them by), nor have I reformatted its text from the numbered paragraphs, which serve in lieu of pagination in internet computer network transmission.

The essay argues, as the whole of this collection in some sense does, that in the late age of print the topography of the text is subverted and reading is design enacted. Thus, the choices a text presents depend upon the complicity of the reader in creating and shaping meaning and narrative. As more people buy and do not read more books than have ever been published before, the book is merely a fleeting, momentarily marketable, physical instantiation of the network. Readers face the task of re-embodying reading as movement, as an action rather than a thing, network out of book.

"Interstitial: Artists' Statements—Giving Way(s) before the Touch" includes two "artist's statements" about my hyperfiction, *afternoon, a story*. The first statement is made up of the four screens I wrote in 1987 to provide directions for readers of this hyperfiction. The ideas of "words which have texture" and yield words attempted to describe a way of reading sensually and imagined the screen as giving way before the touch, lapsing in the rhythm of caress, rather than, as the interactive cinema artist Grahame Weinbren puts it, descending "into the pit of so-called 'multi-media,' with its scenes of unpleasant 'buttons,' 'hot spots,' and 'menus,' [that] leaves no room for the possibility of a loss of self, of desire in relation to the unfolding." Its ideas about closure and continuity seemed unremarkable to me at the time although difficult to express. Even so, six years later, when Robert Coover asked "What is it like to read fiction on a computer screen in hypertext?" in his front-page *New York Times Book Review* article, "Hyperfiction: Novels for the Computer," he ended up quoting from this section of *afternoon* "what is perhaps its most famous line 'There is no simple way to say this.'"

The second statement was prepared as a contribution to one of the "Words on Works" issues pioneer electronic writer Judy Malloy edits for the *Leonardo Electronic News*, published by the International Society for the Arts, Sciences and Technology. It was written at about the same time as "A Feel for Prose," which it borrows from, and in the thrall of a weekend spent partly in conversation with a wonderful young Italian art critic, whom it mentions and who gave my work contexts I had not imagined but nonetheless could see. It asks more questions than it answers. It seems to me as close as I have thus far been able to come to the truth of what, in "The Last Painting," Hélène Cixous calls "the second innocence, the one that comes after knowing, the one that no longer knows, the one that knows how not to know."

"Hypertext Narrative" is another polemic and a palimpsest and, like "Selfish Interaction," lives in doubled time as more an action than a thing. Originally written as a talk for a panel "Hypertext, Narrative, and Consciousness" at the ACM Hypertext '89 meeting in Pittsburgh (where it was presented with accompanying Hypercard screens and distributed in laser-printed samizdat because panel papers were not included in proceedings); it was revised and augmented in 1992 for *Perforations*, the magazine-in-a-box edited by Richard Gess for the Atlanta arts collective "Public Domain." The revised essay includes a new section on virtual reality and interanthologizes the "Dryden Statement," a "minifesto" on interactive fiction written following the first meeting of TINAC, also an (interactive) arts collective founded by Nancy Kaplan, Stuart Moulthrop, John McDaid, and myself. In a nomadic argument that makes its way from Calvino to Eco to Deleuze and Guattari to Hakim Bey, the essay suggests that "the value produced by its readers is constrained by systems that refuse them the centrality of their authorship": "What is at risk is both mind and history."

"What Happens as We Go" expands on a talk I gave for the Museum of Modern Art "Technology in the 90s" series in April 1993 and points toward the victory of the text-as-read envisioned by Calvino. The four parts of its extended subtitle, "Hypertext Contour, Interactive

Cinema, Virtual Reality, and the Interstitial Arts of Jeffrey Shaw and Grahame Weinbren," are not so much strands of one yarn as the quilted stuff that make up an entangled, rhythmic, and successive conversation. It is a conversation that extends through the final three chapters of inversion, subversion, and transgression (though not, of course, in that or any order) regarding polyvocal rhetorics, multiple narratives, exploratory and constructive hypertexts, and hypertext contours. This chapter includes what to my mind is my clearest formulation to date of the slippery, latter notion (a formulation, therefore, that migrates to the concluding chapter that first spawned the notion of contours), and its final section attempts to enact something like contours in an encounter with Weinbren's "interweaving multiple narrative streams." Weinbren and I first encountered each other among the AI and human interface computer scientists, film theorists, filmmakers and others brought together as part of an extraordinary multidisciplinary interactive video research project choreographed by the unsung pioneer in interactive media, Mary Milton, while she was at the Markle Foundation in 1984.

"Hypertextual Rhythms (Part 2)," the first section of the inverted two-part chapter, "The Momentary Advantage of Our Own Awkwardness," snatches surrender from the jaws of the victory of the previous chapter. Presented as a performance piece at the 1992 Modern Language Association (MLA) convention session "Hypertext, Hypermedia: Defining a Fictional Form," it offers an aesthetic of surrender in what David Porush calls "the war between passion and technology," and depends upon Hélène Cixous's sense of the "third self" and the "betweenus we must take care to keep" as a way to sustain the necessary relationship in claiming for constructive hypertext a surrender of writer to reader (the writer who will be).

The second section, called part 1, was solicited as a one-page contribution to a "virtual seminar" on the "bioapparatus" sponsored and published by the Arts Studio at The Banff Centre in Alberta, Canada, in the fall of 1991. Meant as a dialogue with, as much as a critique of, virtual reality, this theoretical narrative argues that "what's missing [from current virtual reality] is the interstitial link, the constant mutability and transport that the word, now electrified... is not used-to (for despite its hegemony it is not conscious as yet; rather, it is we who are used to this in it) but used-for."

The last essay of this collection, "A Feel for Prose: Interstitial Links and the Contours of Hypertext," is my most fully developed theoretical narrative thus far. Though perhaps as many texts have migrated to (and from) this essay as any, the italics here do not indicate borrowings but, rather, the interstitial murmurs of texture and contour. Fashioned more likely in homage than in the manner or intelligence of Julia Kristeva's "Toccata and Fugue for the Foreigner," its first contours are an extended meditation upon the qualities of print and electronic text ("Print stays itself; electronic text replaces itself"), interspersed with narrative-critical interstitials in which critical voices murmur and converse. The latter contours explore the shape of a new theoretical paradigm for hypertext as "a fully coextensive, truly constructive electronic text [that] will present the reader opportunities for capturing the figure of connection at its interstices":

> The evolving contour must be manifest for the reader so she can recognize, resist, appropriate, possess, replace, and deploy the existing contour not just in its logic and nuance but in its plasticity. She should be able to mold and extend the existing structure at each point of replacement and to transform it to her own uses.

In this much, at least, I have been consistent in both minds (though whether consistency is a virtue in the late age of print will be seen in its fool's patterns). In 1986 my first pass statement on multiple hyperfictions in "Selfish Interactions" asked its readers to "suppose a text can anticipate unpredictable variations upon it"; likewise, the 1988 article "Siren Shapes," the first of my pedagogical theoretical narratives suggested that "transformation of knowledge... is the litmus test we should use in judging both exploratory and constructive hypertexts."

In the end, of course, it isn't for me to say whether two minds stuck for some years on a simple idea flap like the fabric of kites caught on a single pole or whether the licking undulation of their fabric in air speaks in the billow of being made up as a bed is. What I can insist is that these patterns of apparent consistency are what Carolyn Guyer calls "*objects* of our lives... inventions that operate somewhat like navigational devices, placemarkers if you will."

"We go on," she tells us, "like waves unsure of the shore, sometimes leaping backwards into the oncoming, but always moving in space-time, always finding someplace between the poles that we invent, shifting, transforming, making ourselves as we go." This is the language of sisters, of the betweenus at the end of time, of hypertext pedagogy and/*as* poetics in the process of re(de)fining each, of minds that dare to hope to penetrate the dark edge of existence comforted by knowing we are not lost to one another.

Part 1: Of Two Minds

Hypertext Contexts

Hypertext and Hypermedia

Hypertext is, before anything else, a visual form. Hypertext embodies information and communications, artistic and affective constructs, and conceptual abstractions alike into symbolic structures made visible on a computer-controlled display. These symbolic structures can then be combined and manipulated by anyone having access to them.

Hypertext structures are often represented as nodes and links: the nodes "contain" text, graphical information, sounds, and so on, connected by the links, which, however, may themselves also contain information or at least be labeled. Depending on the interface metaphor–the way the program visually depicts its information–the distinction between nodes and links is not always represented in hypertext programs, and the contents of the nodes themselves can often function as links. In figure 1 the text node, including the illustration, contains hidden links to further information about Walt Whitman that the reader can access by pressing special keys to display them as they are here. Readers can also follow other explicit links in the "map" window of other nodes and links or use search commands to find other nodes or links, certain topics, or words and phrases.

Borrowing from the conventions of print culture, those who view, combine, or manipulate hypertexts are commonly referred to as readers, while those who create, gather, and arrange hypertexts are called writers. Yet hypertext challenges and, many say, obviates these distinctions. Hypertext readers not only choose the order of what they read but, in doing so, also alter its form by their choices. Also most hypertext systems allow "readers" to add their own material, or links,

This essay first appeared in 1993 as "Hypertext/Hypermedia" in the Encyclopedia of English Studies and Language Arts. Reprinted by permission of Scholastic Inc., ©1994, National Council of Teachers of English.

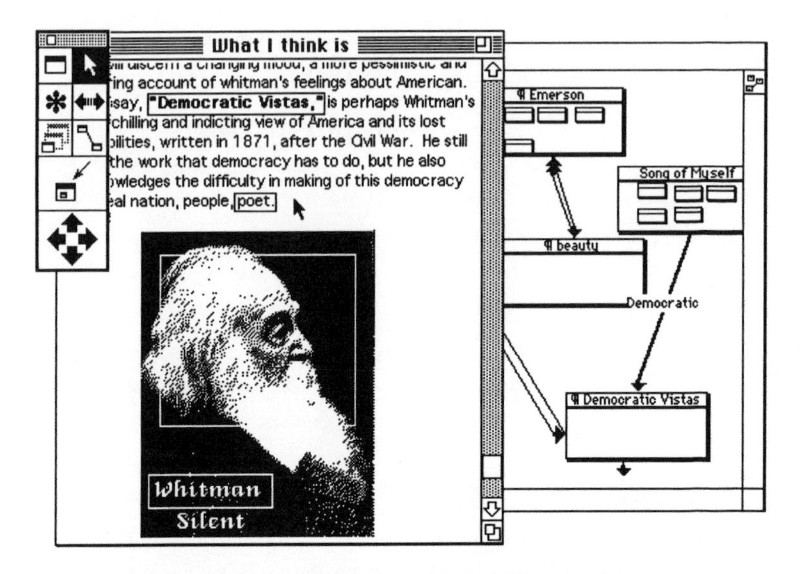

Fig. 1. In the text window at the left, hidden "wireframe" links have been highlighted by the reader, indicating that further information is available about certain words and the illustration of Whitman (note the two white wireframes in the graphic). The map window offers other nodes and links that the reader can access and locates the relationship of the current space to the hypertext. (Whitman Storyspace by Martha Petry and her students at Jackson Community College, used by permission.)

to hypertexts. They thus determine its content for themselves, and often for successive readers and in a very real sense write (or rewrite) hypertexts. This dissolving of distinctions between writer and reader makes hypertext a valuable tool for learning and information management as well as a revolutionary artistic medium. Indeed, some theorists argue that hypertext represents a shift in human consciousness comparable to the shift from orality to print.

Hypertext takes advantage of the computer's ability to retrieve information in any order (random access) and to store it in any form (a hypermedia database). Hypertext enables interaction between viewers of its material and those who created or gathered that material. While such interaction has heretofore largely taken place in successive times and at a distance (asynchronously); simultaneous and multiple (synchronous) systems are already in use in university and industry research settings, especially the World Wide Web (WWW).

Readers access the symbolic structures of a hypertext electronically in an order that they choose. They may select among possibilities already visually represented for them by successive authors of the hypertext or its previous readers, or they may create new choices by discovering them within, or adding them to, the visual organization of the hypertext. Computer scientist H. Van Dyke Parunak (see Berk and Devlin 1991) sees these choices as a variety of artificial intelligence in which humans supply inferences, connections, and natural language processing that computers cannot yet provide. In any case, readers' choices constitute the current state of the hypertext and, in effect, its form. Since even the simplest hypertexts present an enormous number of reading choices, and since the order of presentation always changes with readers' choices, hypertexts can never be adequately represented in print.

Hypertext content, hypermedia, multimedia. Thus far the primary visual symbolic structure of hypertext is language printed to the screen, but, as hypertext theorist Don Byrd suggests, print is a "content, not the form, of electronic media." When hypertext content extends to digitized sound, animation, video, virtual reality, computer networks, databases, etc., it is referred to as *hypermedia.* Hypermedia is electronically rendered in computers and smaller microcomputer-based consumer devices for storing and delivering information and entertainment. These range from "electronic books" and CD-ROM (Compact Disk-Read Only Memory) or CDI (Compact Disk Interactive) devices to "personal assistants," wireless pocket computers connected with telephone and data networks. *Multimedia* is often (confusingly) adopted as a marketing and technical term by computer manufacturers, software developers, publishers, and others to describe both hypermedia content and the hardware or software that embodies it.

A History of Definitions of Hypertext

Like so much of computer science and engineering, hypertext development has been driven by often competing theories of mind. Born of the liberal progressivist agenda of the post-World War II scientific research effort, the development of hypertext has been shaped as much by thinking in cognitive science, literary theory, pedagogy, utopian social thought, and the written and visual arts as it has been by computer science research into human-computer interfaces, knowledge structures, artificial intelligence, database management, and information retrieval. The convergence of such fields in hypertextual development signals not only a momentous cultural shift from print to new electronic media ("this late age of print," as Jay Bolter terms our time) but also, and more important, a corresponding shift in the way humans think. At each stage of the developing consensus about these shifts, new and sometimes contradictory definitions of hypertext have been advanced. Even so, a remarkably consistent line of thinking holds that hypertext in some sense represents the workings of the human mind. (See Conklin 1987, to which this or any other history of hypertext is indebted.)

Human mind operates by association—Bush and Memex. While scholars continue to suggest forebearers of hypertext ranging from the Greeks to Sterne's *Tristram Shandy* and beyond, hypertext (although it was not called such until two decades later) has its roots in a single article ("As We May Think," *Atlantic Monthly* 176 [1945]) by Vannevar Bush, then director of the United States Office of Scientific Research and Development. There Bush describes the "memex," a mechanical, microfiche-based, see-through desktop, "a device in which an individual stores . . . books records and communications . . . an enlarged intimate supplement to . . . memory." The memex, better suited to electronic and digital than to gear-driven information retrieval, was never built. Yet Bush's insistence that "the human mind . . . operates by association . . . in accordance with some intricate web of trails carried by the cells of the brain" has influenced all subsequent hypertext theory and development, both in its fundamental attention to cognition and in its conceptual framework and vocabulary.

Augmenting human minds—Engelbart and NLS/Augment. What Vannevar Bush could only imagine, Douglas Engelbart not only imagined but also designed and built. In the process he invented or first put to serious use fundamentals of computer interaction, writing, and networking, including word processing, outlining, windows, electronic mail, computer conferencing, collaborative authorship, and—not last—the mouse. His 1962 essay, "A Conceptual Framework for the Augmentation of Man's Intellect," led to his development of a full-blown prototype hypertext system NLS (oNLineSystem) at the Science Research Institute in 1968. Renamed AUGMENT, this system was the beginning of Englebart's lifelong exploration into what he terms "a co-evolutionary process—new knowledge processes and new tools evolving together in real working environments" (Berk and Devlin 1991).

Mind as intertwingled "Docuverse"—Nelson and Xanadu. The last of the founding trinity of hypertext was also its baptizer. Provocateur-humanist, Theodor Holm (Ted) Nelson provided a utopian spirit and encompassing vision (as well as the coinage, hypertext, in the 1960s) to the new world of the mind imagined by Bush and established by Engelbart. Nelson is most likely also responsible for what George Landow has termed the "bizarrely celebratory" quality of writing about and with hypertext. Proceeding from a belief that "literature is an ongoing system of interconnecting documents," Nelson (1990) has joyfully and wittily prophesied and proselytized the coming of a "docuverse" not unlike the linked world (as) text of contemporary computer networks. For decades Nelson has promised to embody the docuverse in his own Xanadu system, an ongoing design prospectus geared toward establishment of a peaceable kingdom of "intertwingled" and computerized text on earth (see Nelson 1987b).

Writing the mind—Bolter and Storyspace. It took the oddly fitting combination of someone who is both a classicist and computer scientist to place hypertext both in its historical perspective as the successor to print and at the center of cultural scrutiny. Jay David Bolter's sometimes controversial book *Writing Space* (1991) challenges assumptions of contemporary literary theory and computer science alike while making the case for hypertext as a new writing technology. "Electronic writing is both a visual and verbal description," says Bolter, "not the writing of a place, but rather a writing *with* places, spatially realized topics . . . signs and structures on the computer screen that have no easy equivalent in speech" (25). For Bolter, hypertext's "electronic symbols . . . seem to be an extension of a network of ideas in the mind itself." Storyspace, the hypertext system Bolter and I developed with John B. Smith, embodies Bolter's view that the "topographical" writing of hypertext "reflects the mind as a web of verbal and visual elements in a conceptual space."

Visions of Hypermedia

Hypertext has been called the revenge of the text on television since under its sway the screen image becomes subject to the laws of syntax, allusion, and association, which characterize written language. Print literally gives way on hypermedia screens to digitized sound, animation, video, virtual reality, and computer networks or databases that are linked to it. Thus, images can be "read" as texts, and vice versa. Any hypertext holds the prospect of representing on the screen the sights, sounds, and experience of movement through virtual worlds that language previously only evoked in the imagination.

Engelbart's mouse discovers virtual reality. By linking movements of the human hand with operations upon symbolic structures (words, windows, icons, etc.) on the computer screen, the electromechanical mouse, the now familiar interface device for which Douglas Engelbart holds a patent, opened the way to virtual reality (VR). Virtual reality enacts continuously linked simulations within computer environments in response to body movements interpreted by the computer through input gleaned from "data gloves," special goggles or helmets, and the like. While fully represented virtual reality as yet remains a costly research endeavor and entertainment enterprise, hypermedia simulations are regularly used in education, training, marketing, and entertainment. The design and storyboarding of these simulations likewise involves the use of hypermedia editing, presentation, or desktop video tools. Graphical, "object-oriented" interface features, borrowed from CAD (Computer Aided Design) environments, script and create virtual worlds.

Continents of knowledge from FRESS to Intermedia. Windowed and interconnected worlds of knowledge—now a commonplace of hypermedia (especially virtual reality)—owe a significant debt to a series of hypermedia systems developed at Brown University between 1968 and 1991. Hypertext pioneer Andries Van Dam, a graphics specialist, explored the development of multiwindow systems, developing the Hypertext Editing System (HES) in 1968, quickly followed by FRESS (File Retrieval and Editing System). Not designed primarily for teaching, FRESS was almost immediately used as the vehicle for networked instruction in a poetry course led by the critic Robert Scholes. FRESS was followed by Intermedia, perhaps the most influential educational hypertext system in the brief history of hypertext. Members of Van Dam's research team, including Norm Meyrowitz, Nicole Yankelovich, and Paul Kahn, were joined by Brown faculty members, most notably English professor and hypertext theorist George P. Landow, in the development of Intermedia. *Webs*, Van Dam's term for sets of hypertext links, were incorporated into vast corpora ("continents of knowledge") of primary and secondary materials (including student commentaries) for courses in English, Biology, and other disciplines. Landow's "concept maps" joined interactive graphical timelines and web views as visual symbolic structures available to Intermedia learners. With Intermedia's demise in 1991, Landow transported his Dickens and *In Memoriam* (Tennyson) hypertext webs to Storyspace, publishing them as electronic textbooks and learning environments.

Picnics in the PARC: Notecards, Hypercard, and their clones. Engelbart's and others' early work in large-scale, networked, collaborative hypertext systems continued in industrial, corporate, as well as university research settings through the 1970s into the mid-1980s. At that time research systems came to market, beginning with KMS, a version of the Carnegie-Mellon research system, ZOG, in 1983, followed in 1985 by Peter Brown's, Guide, the first commercial hypertext system for personal computers. The predominant current metaphor for hypermedia authoring—cards with buttons—emerged with the development of Notecards at Xerox PARC (Palo Alto Research Center) by Frank Halasz, Thomas Moran, and Randall Trigg. Notecards spawned important research into hypermedia screen design, collaborative writing, and hypertext navigation, much of which continues today in the Aquanet hypermedia system. Both the Macintosh (Mac) computer and the slyly named Hypercard software (which preempted as much as it popularized hypertext) were strongly influenced by the graphical interfaces and the hypertext metaphors developed at Xerox PARC. Likewise, most multimedia authoring systems owe their genesis to Notecards.

Current Controversies

Like any cultural shift, hypertext has been accompanied by controversy and contention, much of it having to do with disputes about the nature of mind and especially who holds the power over symbolic structures. Some see hypertext as another way of reading; some see it as a new way of knowing—yet what is read and who should know what are both in dispute.

Alternate views and cautionary tales—or the great "and/and/and . . ." Taking up the utopian promise of Bush, Engelbart, and Nelson, hypertext theorists and writers have argued that its new ways of knowing offer, in John McDaid's words, "to shape a digital culture of empowerment and difference" (in Berk and Devlin 1991, 457). Jane Yellowlees Douglas (1991) sees hypertext as accommodating multiplicity by "rejecting the objective paradigm of reality as the great 'either/or' and embracing, instead, the 'and/and/and'" (125). I argue for a "constructive hypertext" as "a structure for what does not yet exist" (1988, 12).

Reconceiving hypertext. While not rejecting this promise, Stuart Moulthrop, in a prolific series of essays (see Berk and Devlin 1991 and Hawisher and Selfe 1991), has argued that hypertext could be preempted by the "military infotainment establishment" or offered as a diversion to a dissatisfied society in lieu of real access or power. Teacher/theorist Nancy Kaplan asks "how... hypertext and hypermedia structures... work, rhetorically and ideologically": "Whose interests and visions, whose realities, do... [they] serve?" (Hawisher and Selfe 1991). Other feminist critics and hyperfiction writers describe open, inclusive structures. Carolyn Guyer and Martha Petry (1991b, 82) call for a "deconstruction of priority" in which "connection itself [is] a figure against the ground of writing." Catherine Smith's (1991, 125) "field description of knowledge... would include the nature and dynamics of 'the inner life,' or *affective processes*: forgetting and denying as well as remembering and recognizing associations, rejecting as well as acknowledging connections."

"Hypertext engineering" versus the docuverse. So-called real world concerns of preparing "industrial strength hypertext" (see Parunak 1991) lead critics such as discourse theorist Davida Charney to consider the suitability of hypertext "for transmitting verbal information." Technical requirements of hypertext documentation for aircraft carriers or aircraft assembly hangars occupies hypertext research literature (see Nielsen 1990). "Hypertext engineer" Robert Glushko (1989) emphasizes "a disciplined approach... [resting] on task analysis, document analysis and careful consideration of... technical writing and information retrieval disciplines." Charney argues that hypertexts may not "meet the needs of readers and writers who depend on the text to help... sequence the flow of ideas through focal attention" (1994, 213).

Lost (and found) in hyperspace: breadcrumbs, compasses, and cognitive prostheses. Such concerns with "cognitive overhead" in hypertext have generated theory and computer interfaces alike. Fittingly, much of both has come from hypertext publisher and researcher Mark Bernstein, whose Eastgate Systems pioneered the publication of interactive hypertext fictions and nonfictions for both the trade and academe. Bernstein's "breadcrumbs" (Hansel and Gretel-like markers for hypertext trails), compasses, and link apprentices ("swift, friendly, but dumb" search programs) are examples of what Patricia Wright (1991) has termed "cognitive prostheses" against temporary overload—or being "lost in hyperspace." Like many researchers, however, Bernstein (1991a) has recently begun to look at how "hypertexts can and do exploit disorientation" and to mark distinctive contours—or what philosopher David Kolb calls the "new figures"—of hypertext writing.

Copyright and its discontents: the restaurant at the end of the universe. Hypertext calls into question the several-hundred-year-old copyright system. Computer networks (or hard disks in offices, classrooms, and homes worldwide) belie attorney Stephen L. Haynes's confident assertion (in Berk and Devlin 1991) that because "hypertext and hypermedia look and act differently bears little on the intellectual property factors that come into play." Like diners in the restaurant of Douglas Adams's novel *The Hitchhiker's Guide to the Galaxy*, authors and publishers in the late age of print watch their universe bang into being and implode into a black hole with each revolution. Legal scholars and industry analysts agree that something must be done, but few real solutions seem evident. Ted Nelson's earliest Xanadu scheme included a fully articulated intellectual property system, yet copyright attorney Pamela Samuelson and Robert Glushko (1991) fault its "model of author and user behavior and the adequacy of [its] financial incentives for authors." Copyright attorney Patricia Lyons suggests (in Berk and Devlin 1991) that, as "the manipulation of text in order to abstract what may be termed 'knowledge'" increases, "it is evident that notions of 'fair use' developed in a print world may no longer be relevant." Meanwhile, however, for wont of a system, the publishing industry relies on market forces and the economy of scale to secure their revenues, if not rights, through "repurposing" of printed text and audiovisual materials into massive CD-ROM collections, "Expanded Books," and electronically distributed network publications.

Current Uses and Future Prospects

The Web of "Real-world" and "Real-time" hypertexts. The promise of hypertext is increasingly bound into its reality, its theory tested by its practice. Scholarly, pedagogical, scientific, business, engineering, medical, and legal hypertexts are in wide use. Literary hypertexts especially have received critical and media attention, challenging notions of textual closure, authorship, and reader response. While educational hypermedia is increasingly available in various disciplines for both K–12 and higher education, much of it is still of dubious value, both since it largely consists of repurposed textbook and audiovisual materials and since most schools lack the equipment necessary for wide-scale interaction. Even so, educational hypermedia such as the ABC Interactive series and Robyn Miller's experiential narratives *Manhole* (1988) and *Cosmic Osmo* (1989) offer compelling simulations and exploratory spaces, while the Computer-Supported Intentional Learning Environments (CSILE) developed by Marlene Scardemelia and others encourage "making knowledge-construction activities overt, maintaining attention to learning goals . . . , providing process-relevant feedback, and giving students responsibility for contributing to each other's learning."

Mosaic, the hypertext browser for the internet, offers a version of Nelson's docuverse enabling scholarly, artistic, and commercial interaction, collaboration, "infotainment," and publication throughout the world. Mosaic was developed by the National Center for Supercomputing Applications (NCSA) at the University of Illinois as a public domain visual interface to the World Wide Web. The WWW project, which originated at CERN (the European Laboratory for Particle Physics in Switzerland) has experienced explosive growth in the mid-1990s following the development of Mosaic. Thousands of sites and collections—including libraries, scientific, technical, and artistic journals, academic and commercial publishers, art galleries, video archives and so on—are accessed through "Uniform Resource Locators" (URLs), which provide graphical addresses for shared expression and information.

How these educational, scientific, and artistic developments will shape future culture is obviously uncertain. Virtual reality, synchronous conferencing, and networked hypertext systems will emerge from university and industry research labs into common use at much the same time that high-definition television, optical fiber and satellite networks, and miniaturized computers reach consumer markets. Whether such networks can bridge widening gulfs between developed and less developed countries and economic and social classes will assume paramount importance in browsing through continents of knowledge and mapping the topological multiverse.

What I Really Wanted to Do I Thought

In 1982 I had published a prize-winning small press novel, and, like any novelist, I knew that it was an awful thing to have to type successive drafts of a book-length manuscript—and, a worse thing to discover, midway in the last draft, that something on page 265 would work wonderfully on page 7. I believed that a word processor would allow me an option for making such changes beyond the increasingly difficult task of coercing wife and friends to help me again retype the final manuscript.

Immediately, I discovered that what I *really* wanted to do was something else. What I really wanted to do, I discovered, was not merely to move a paragraph from page 265 to page 7 but to do so almost endlessly. I wanted, quite simply, to write a novel that would change in successive readings and to make those changing versions according to the connections that I had for some time naturally discovered in the process of writing and that I wanted my readers to share. In my eyes, paragraphs on many different pages could just as well go with paragraphs on many other pages, although with different effects and for different purposes. All that kept me from doing so was the fact that, in print at least, one paragraph inevitably follows another. It seemed to me that if I, as author, could use a computer to move paragraphs about, it wouldn't take much to let readers do so according to some scheme I had predetermined. What's more, I knew that my writing students could benefit from the ability to recognize and manipulate internal connections in their writing.

By that time I had begun to read numerous computer magazines and sometimes even begun to understand an occasional word or sentence in the accompanying ads. At any rate, I found in some obscure computer magazine a thoroughly frightening article about a woman, Natalie Dehn, at the Yale Artificial Intelligence Lab who was trying to teach computers to write novels. Her face seemed kind, so I wrote her a letter and asked her to tell me, please, where I could buy a computer program that would let me write novels that changed every time someone read them. She wrote me back an eight-page letter that explained how this was impossible and would be for twenty years; she argued—vociferously and much more ably than most literary critics I know—for the virtues of linearity in prose fiction and for the aesthetic function of constrained choices in imagination.

I began a long correspondence with Natalie, which resulted in my invitation to Yale and, more important, her suggestion that I contact Jay Bolter, who she said had previously been a visiting fellow there, was "also crazy," and was interested in similar questions. My correspondence with Jay led to foundation funding of my sabbatical and initial work on our program. We and his colleague John Smith went on to create Storyspace, which in the spring of 1987 enabled me to write a hyperfiction, *afternoon, a story,* that changes every time you read it.

We had generally completed the underlying functionality of the program before we heard the term *hypertext* or read Ted Nelson's *Literary Machines* (1990). Yet from the earliest point in our collaboration we progressively found ourselves in contact with the tradition of hypermedia studies, beginning with Bush and Engelbart and continuing to Nelson. That tradition of scholarship and active collaboration existed as something of an iceberg—or, more aptly, like some huge octopus with only its eye above water but with submerged tentacles reaching almost everywhere around us, including pedagogy, linguistics, cognitive science, literature, physics, database theory, classics, media studies, medicine, and so on. Because it was a tradition concerned with links and interrelationships, it observed no intellectual boundaries.

It was *the* computer—with its essentially integrative, relational, recursive, projective, but stubbornly, even stupidly, logical nature—that seemed, first, to enable us to join this previously invisible learning community and, second, to join in active dialogue within that community. As we appropriated computers to our uses and modeled complex understandings upon a foundation of low-level concerns, we found ourselves in dialogue with others who, though they proceeded from much different disciplines, shared a common process of tool building and intellection with us.

I knew of, and for years had given lip service to, the interdisciplinary nature of my professional life. I had done my work at the Iowa Writers Workshop and so was in touch with a widespread and active artistic and learning community as well as the scholarly and critical community that Sherman Paul had introduced me to. I had trained myself as a composition theorist in the line of fire, as chair of a community college English department, and knew that field through research, practice, conferences, and anxiety attacks. Not only did each of these domains interact with one another; they also actively espoused essentially interdisciplinary stances.

Despite all this, the problem was that, if we talked at all, my colleagues and I more often than not spent more time talking among ourselves about interdisciplinary learning than we did putting it into practice. It was like a floating crap game in which we were eager to play, if somebody would only tell us where it was and pass us the dice. I had cochaired a writing across the curriculum project and had team-taught a graduate professional development course for our faculty; I chose to teach in a developmental education program that was interdisciplinary by nature. Still, no dice.

Then I bought the computer, and there were the dice. The technological shaman for my rite of passage was Howie Becker, the well-known sociologist and jazz pianist, who had an Apple II, wrote letters with multiple fonts before there was a Mac, and shared software, ideas, and dreams. Over the years we have competed to outdo each other in having smarter software, bigger computers, and better dreams.

I continued to shoot the dice during my year at Yale. The whole AI project was alive with people who observed few boundaries but cared for my issues as Natalie Dehn did. Though for a time I had to learn to identify myself as "a story generation person" to an uncomprehending few, who apparently hadn't heard the term *novelist*, I sat as a novelist as much as a would-be software designer in the presence of Peck's Bad Boy of AI, the inimitable Roger Schank, whose only requirements for my presence there were that "1) you don't get in anybody's way and 2) you swear that you won't write about me." I kept half these promises to this day.

To a novelist the people at Yale seemed to share a conviction that learning is the outcome of the stories we tell ourselves as we come to know new things, no matter how abstracted, technical, or objective our conception of those things. These people *cared* about creativity and learning because they knew how difficult it was to model the least instance of it. They shared Schank's vision that thought discloses itself in connections rather than in substance and that learning is characterized by the activity of explanation.

In the midst of this experience I came to understand that computer tools enable "fervent errors," the kinds of errors Thoreau called extravagance, i.e., literally, walking off the path. Computing is a wonderfully liberating environment for fostering error—especially to an English teacher used to fighting students' expectations that he opposes rather than encourages errors. All programs have bugs; many have system bombs or fatal errors; some destroy boot blocs, VTOCS, and letters to Aunt Ida. Computer dictionaries uniformly stink, syntax checkers are worse, spreadsheet formulas veer into circular references, somewhat trained logicians (such as I) find themselves writing conditional branches on A or not-A. It was common for researchers in Schank's group to conclude a private demo of an AI program able to understand dauntingly technical or abstruse samples of what is quaintly called natural language by asking whether you would like to see the program "urp and crash" on a laundry list or a love note.

By the end of my Yale year Bolter and I had a (mostly) working version of Storyspace, which, in January 1986, I began to use with developmental writing (what used to be called remedial or college preparatory writing) students and soon after also began to use with my creative writing students. Since then my colleagues and our students have used it in literature and humanities (cultural studies) courses, technical writing, business writing, and so on, while uses elsewhere run the gamut from sociological and anthropological field notes to political strategy and planning to earth sciences to historical studies and, of course, the extraordinary Lanestedt/Landow webs from Brown University.

Sometimes I am surprised by what others do with Storyspace, though never as surprised as Jay Bolter, who unfailingly exclaims, "Don't tell me people are using it for important work!" when he hears of any news. I am, however, almost always caught unprepared when our work is described as visionary. I think of my father, who once, when I called him excitedly to report on an all-expenses-paid trip to a foundation-sponsored thinktank session at a West Coast spa that put me among filmmakers, computer interface specialists, and others, said "Aren't you lucky to be able to be in the room with so many experts."

Oddly enough, my best understanding of what it might mean to be visionary proceeds from a homily Roger Schank gave to his graduate seminar during the year I visited his Yale project. As best I could tell, this particular seminar largely consisted of Roger, dressed in any of an infinite number of genuine rugby shirts, reading from the daily *New York Times* while commenting on human cognition as it expressed itself in goals, plans, explanations, and certain foolish opinions he goaded the graduate students into expressing or that they blurted out on their own. All in all it was a tour de force that everyone seemed to enjoy, if only for its sense of apocalyptic promise and mordant wit. I enjoyed Schank's working-class bravado, flashing eyes, and the moment when, right as rain, the hour struck and throughout the room the digital watches of the best young minds in computer science beeped one after another in no apparent pattern or synchronicity. In one such session, prompted by something I can no longer recall but surely not anything that seemed weighty at the instant, Roger leaned back and addressed the students ex cathedra and outside the veil of the great gray lady, giving advice directly and with an affection and fire I still recall with genuine respect and affection.

"In political matters," he said, "go the way of most resistance, because what you believe you likely haven't thought through. But in the way of intellectual matters and your own research, go the way of least resistance, for what comes easily to you will likely be hard for anyone else to understand and will probably represent a real discovery once you can explain it to them."

Like a novel that changes with every reading.

Part 2: Siren Shapes

Hypertext Pedagogy

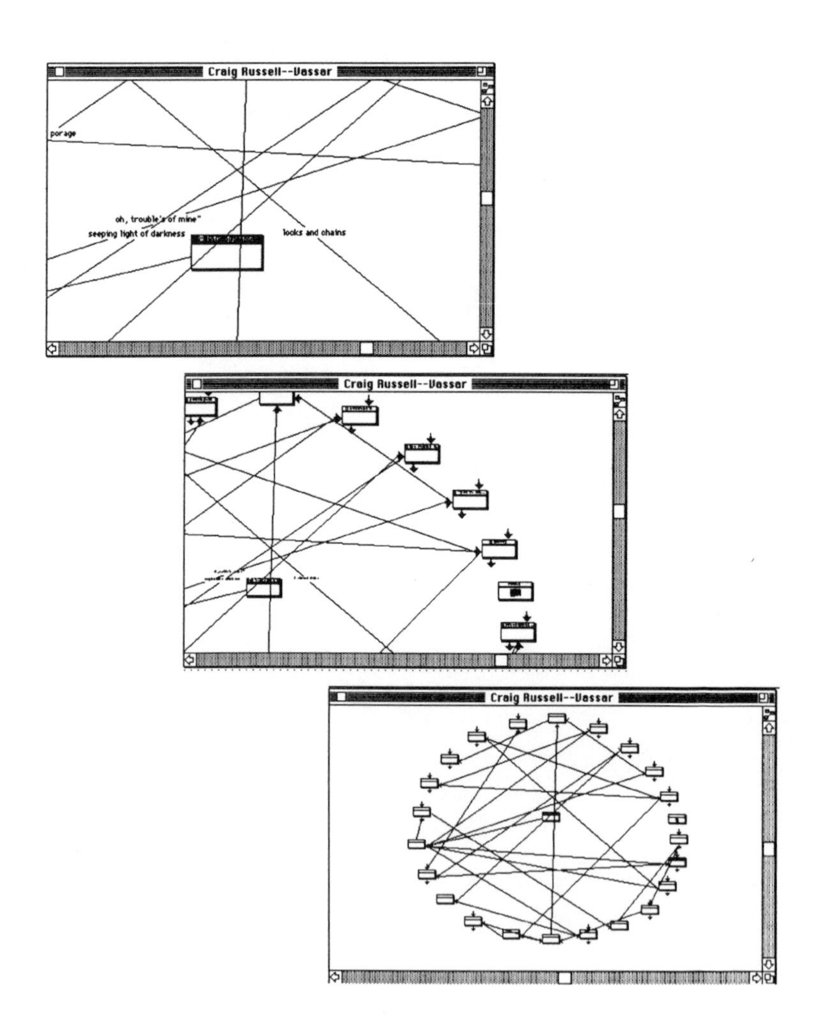

From a heliotropic reading of Erin Mouré and Lucille Clifton's poems in Storyspace by Craig Russell, Vassar '96

Siren Shapes: Exploratory and Constructive Hypertexts

Hypertext and hypermedia are increasingly perceived as instances of a cardinal technology, i.e., tools for working at traditional tasks that have the effect of changing the tasks themselves. Yet a fairly common reaction to hypertext and hypermedia systems in product reviews, in technical literature, and among everyday users of these tools is the one expressed by Jeffrey Conklin in his still definitive 1987 article, "One must work in current hypertext environments for a while for the collection of features to coalesce into a useful tool" (18).

This is a kind way to say that you have to figure out what to do with these things. I have spent much of the last four years figuring out exactly that. As a codeveloper of Storyspace, I have approached what to do with these things as a design question; as a fiction writer seeking to work in a new medium, I have approached it as an artistic question and, as a teacher, a practical, pedagogic, and sometimes a political one.

My colleagues and I have spent years using hypertext tools in a variety of settings within a comprehensive community college. Storyspace has been used not only as a day-to-day writing and thinking tool but also in chemistry, nursing, technical writing, creative writing, literature, and developmental reading and writing courses. Users include faculty, community professionals, creative writers, and a wide range of traditional and nontraditional students, from high school aged to senior citizens.

Our experience in using Storyspace predates the release of both Guide and Hypercard; it began with fairly unstable developmental versions and continues as Storyspace is shaped for commercial release. It has helped us test our notions about learning and has informed our work in designing hypertext learning tools using both Storyspace and Hypercard. To some extent, then, given this kind of applied design and testing, I have also approached the question of what to do with hypertexts as a research question.

Given all these claimed approaches, it would seem that I ought to have gotten somewhere; I will try to say where that might be. I will even take the obligatory stab at explaining what hypertext is as well as what it might become. (Hereafter I will use the term *hypertext* where *hypermedia* would do as well, since nearly all hypertext systems involve other media, and I know of no hypermedia systems that use no text.) Before that, however, I want to discuss—briefly and, in the hypertext spirit, idiosyncratically—some design issues, artistic issues, and ultimately practical, pedagogic, and political issues that we too often ignore at our peril. In doing so, I want to offer a somewhat polemical (and, for a hypertext developer, arguably self-serving) description of some of these perils as well as of the promises that accompany them.

I

Indeed, hypertext tools offer the promise of adapting themselves to fundamental cognitive skills that experts routinely, subtly, and self-consciously apply in accomplishing intellectual tasks. Moreover, hypertext tools promise to unlock these skills for novice learners and to empower and enfranchise their learning. Ironically, however, our ability to deliver upon these promises may be imperiled in the short run by many of the same factors that make this technology so promising.

For instance, the ready adaptability of these tools to more traditional uses is especially compelling given the technological frosting they so easily spread upon stale cake. This disincentive to change is in no way novel, either in the long history of cardinal technologies or, especially, in the short history of microcomputers in education. The adaptability of Hypercard, for instance, makes it easy to "author" educational software that merely redistributes the command lines of the worst kinds of supposedly interactive, "drills and skills," CAI software into gaily embossed buttons and peekaboo card fields. Like the Applesoft Basic revolution in educational software that preceded it and that it so clearly resembles, the Hypercard revolution requires us to rely upon skeptical eyes, keep a shrewd ear open to word-of-mouth (or word-of-network) advice, and exercise a cool hand (and quick delete finger) in choosing among a burgeoning list of titles.

Because the price is right (again like the Applesoft revolution, much of this software is "shareware" or otherwise relatively inexpensive), it is likely that the potential benefits outweigh nearly all the short run perils, save perhaps the most crucial one. The peril of overpromising threatens not just to sap the resilience of educators, who must wade through the dross and justify the costs. It also threatens the credibility and creativity of innovators, who find themselves having to disaffiliate and differentiate before they can discover. Avoiding the peril of overpromising, I will argue, depends upon our ability to distinguish between what I call exploratory and constructive uses of hypertext as a learning tool and our willingness to pursue and encourage the development of both.

By exploratory use, I mean to describe the increasingly familiar use of hypertext as a delivery or presentational technology, as Guide and Hypercard are currently most often used. Exploratory hypertexts encourage and enable an audience (*users* and *readers* are inadequate terms here) to control the transformation of a body of information to meet its needs and interests. This transformation should include a capability to create, change, and recover particular encounters with the body of knowledge, maintaining these encounters as versions of the material, i.e., trails, paths, webs, notebooks, etc.

The hypertext audience should also be able to readily understand the elements that make up a particular body of knowledge, plot their progress through these elements, and locate them at will. These so-called navigational capabilities should be present both within the organizational structure of the hypertext and from the perspective of the particular versions of it that the audience creates. At least in the short run, and especially in educational hypertexts, the audience should be able to view alternative visual representations of the structure of the corpus and, in some sense, be able to differentiate the unique organizational schemes of hypertext from the more conventional organizations of print and other media. Ideally, an exploratory hypertext should enable its audience members to view and test alternative organizational structures of their own and perhaps compare their own structures of thought with hypertext and traditional ones.

By constructive use, I mean to describe a much less familiar use of hypertext as an invention or analytic tool, such as the uses we have designed for and made of Storyspace. These are also the uses to which outline processors and their offspring, Personal Information Managers (PIMs) such as Agenda and Grandview, have been put. For that matter, these are also the uses that have sometimes been forced upon word processors, spreadsheets, and databases. Just as exploratory hypertexts are designed for audiences, constructive hypertexts are designed for what Jane Yellowlees Douglas (1987) has, following Barthes, termed "scriptor[s]." Scriptors use constructive hypertexts to develop a body of information that they map according to their needs, their interests, and the transformations they discover as they invent, gather, and act upon that information. More than with exploratory hypertexts, constructive hypertexts require a capability to act: to create, change, and recover particular encounters within the developing body of knowledge. These encounters, like those in exploratory hypertexts, are maintained as trails, paths, webs, or notebooks, but they are versions of what they are becoming, a structure for what does not yet exist. Constructive hypertexts require visual representations of the knowledge they develop. They are, in Jay Bolter's phrase, "topographic writing."

Like the audience of exploratory hypertexts, scriptors should be able to readily understand the elements that make up a developing body of knowledge, plot their progress through these elements, chart new ones, and locate them at will. These so-called navigational capabilities should be present within both the developing organizational structure of the hypertext and within the particular versions of it that the scriptor discovers. Scriptors must be able not merely to view but also to manipulate alternative visual representations of the invented structure of the hypertext, switching between them with a minimum of intellectual effort, or what Conklin has termed "cognitive overhead." They must be able to differentiate and label unique organizational schemes of hypertext, plotting them against more conventional organizations of print and other media, and generating either kind of organization at will. A constructive hypertext should be a tool for inventing, discovering, viewing and testing multiple, alternative organizational structures as well as a tool for comparing these structures of thought with more traditional ones and transforming one into the other.

Transformation of knowledge, I would suggest, is the litmus test we should use in judging both exploratory and constructive hypertexts. It is a critical test in judging whether courseware authored with hypertext tools engages learners in looking at material in new ways or merely looks like a new way of learning. In many ways, of course, this kind of test is not new to us. Understanding, plotting, navigating and recreating knowledge structures is the essence of learning. As current critical thinking across the curriculum attests, however, we are less and less certain of our ability to convey these skills.

This uncertainty, coupled with the novelty and ease of authoring with hypertext tools, raises the peril of overpromising. Too often, despite our inbred intellectual skepticism and knowledge about so-called free lunches, we in education have approached technology with what might be called a hunger for automaticity. Both well and poorly designed exploratory hypertexts feed this hunger. It is easy to think that, because learners can move through a body of knowledge in new ways, they know where they are going. We long for a learning machine and think that the computer will do, even though we know it did not the first time around (and that the SRA carrel, with its "programmed learning" buttons, did not before it).

Poorly designed exploratory hypertexts often involve a second or third coming of the learning machine. They assuage the hunger for automaticity with the full-bellied inertia of tradition. We know the bulk of this stuff; we have chewed on it for years. It is easy to think that, because learners can move through a body of knowledge in new ways, we know where they are going.

Yet neither we nor they know as much about learning as either party would think. Paradoxically, we know too much about learning to make such claims. It is a fair bet to say that our age is at least as likely to be known in the future as the Age of Learning as it is by the ordained cliché the Information Age. The body of knowledge about learning in psychology, cognitive science, neurophysiology, artificial intelligence, and so on, would itself make a rich exploratory hypertext.

The individual versions we might create as we wend our way through this corpus would likely share certain sets of contrary attributes. Learning is multiple yet integrative, difficult yet universal, not easily schematized yet apparently systematic, inherently personal and yet socially manifested, and so on.

These contraries provide cautionary measures against which to judge exploratory hypertexts as learning tools. They also introduce and outline the promise of what I have termed constructive hypertexts. Every well-designed exploratory hypertext proceeds from a constructive hypertext created by its author or team of authors. The transformation of knowledge that an audience works upon an exploratory hypertext in some important sense parallels and rehearses the prior constructive encounter of its authors' associative thought processes. The word *parallel*, however, in this instance is almost certainly metaphoric—and inadequate at that, for this process is sometimes orthogonal, sometimes congruent, sometimes isomorphic, but always, in some important sense, anticipatory. The authors and audience of hypertexts share a transforming interrelationship. They are, to use an overused term, colearners. Even the most transparent exploratory hypertexts involve a shared process of mapping this interrelationship, while constructive hypertexts make the transparent mapping visible, active, and personal.

The importance of the *process* of associative thinking is suggested in a preliminary evaluation of the Intermedia project at Brown University. Discussing George Landow's ground-breaking work in designing the Intermedia English 32 course, Institute for Research in Information and Scholarship (IRIS) investigators reported an "unintended consequence of [their] research, discovered when Professor Landow was forced to teach... before the [Intermedia] workstations... were ready... was that he changed the way he organized his course": "As a result he felt that students grasped pluralistic reasoning styles far better than in previous years (Beeman et al. 1987)" The authors noted that "students were also far more satisfied with the course than in previous years" and presented data showing interesting shifts in students' evaluation of both the amount learned in the class and their overall evaluation of the course.

Both sets of evaluations rose significantly when the course was offered before the workstations and exploratory hypertexts were available. When the course was offered with the workstations, both sets largely maintained gains but, interestingly, fell off to previous levels (or beyond) among students who either rated their learning as less or the class worse than other classes at Brown. Meanwhile, high-end ratings fell off enough to make them of interest.

At least one implication of these shifts is worthy of further research and, more pertinently, suggests the potential for constructive hypertexts as instruments of learning. Landow's reorganization of his course might be said to have mirrored his associative (or pluralistic) thought processes in creating a constructive hypertext, i.e., the design for the exploratory hypertext, English 32. As a distinguished scholar and critic, he certainly possessed these associative, pluralistic thought processes well before he set out to represent them in a hypertext. Yet the benefits of doing so were dramatically perceptible to his students. It would seem to follow that these same benefits ought to be extended to learners themselves, especially if further investigation of such "unintended consequence[s]" yields evidence that exploratory hypertexts produce benefits to instruction that are short-lived or, at least, subject to degradation.

It would only be stirring the hunger for automaticity to suppose that, simply by the instrument of creating constructive hypertexts, students could match the prowess of a practiced scholar in associative thinking. Yet it would be equally unwise, and something overpromising, to suggest that students could gain that prowess by simply exploring a scholar's representation of it. Landow's Intermedia project, to be sure, does not make either mistake; while largely a vehicle for exploratory hypertexts, it provides powerful constructive tools for learners to use in transforming bodies of knowledge.

The importance of an anticipatory, transforming interrelationship among colearners may perhaps not represent a novel contribution to our understanding of learning, but hypertext, as a cardinal technology, does offer a novel environment for enabling and exploiting that interrelationship. Constructive hypertexts renew an ancient promise, one that would have us know ourselves and become authors of our learning.

This author's role in this transforming interrelationship is like that of Jane Yellowlees Douglas's "scriptor for the potential experiences of... readers." While Douglas concerns herself with hypertext interactive fiction, her insights into the importance of authorial intention hold true for more expository, exploratory hypertexts, in which scriptor-learners intentionally become their own readers. "The yields we select, the defaults we discover, influence our understanding of the contents of the text we read," says Douglas; "in most cases, we realize that we have, somewhat unwittingly, made certain interpretive or navigational decisions based upon our apprehension of authorial intention." When the author is oneself, apprehending authorial intention becomes a discovery of one's own distinctive structures of thought.

Douglas focuses upon the literary-critical implications of this apparent and awkward resurrection of authorial intention as a "subject" of literary texts. Other hypertext theorists such as Diane Pelkus Balestri, Jay Bolter, and Frank Halasz address the relationship between authorial intention and structures of thought more directly. In the February 1988 issue of *Academic Computing* Balestri discusses the "contractibility" of what she calls "softcopy" (echoing a term that, Ted Nelson informs me, he coined in the mid-1960s). In Balestri's usage, softcopy—i.e., text on the screen rather than in print—leads to an understanding of "text as having patterns, often multiple patterns for a single text; [and] defines coherence in terms of linkages among parts of a text." Hypertext, she suggests, "unlike softcopy, changes the relationship between writer and reader. The reader becomes a collaborator, constructing and reconstructing the text, choosing his own path through it" (17).

The differences Balestri sees here, between the coherent patterns of links among parts of a text and the constructed patterns that a reader makes, might prove less a distinction between softcopy and hypertext than another description of the interrelationship between scriptor and audience. In any case, Balestri points to the need for training hypertext audiences in the new habits of thought necessary to perceive coherence in patterns and links as well as to generate coherent patterns and links of their own.

These concerns are, as I have noted, not much different from the concerns we bundle under the rubric of critical thinking or the general category of learning. Constructive hypertexts address these concerns in a more conscious way than exploratory hypertexts. They enable audiences of expert and novice readers alike to act as scriptors and to focus upon the discovery of coherent structures and linkages, and, most important, to use a full range of cognitive skills, especially visual ones, to discover new structures and linkages. Balestri's notion of softcopy invites us to consider how coherence can be (and is) visually represented and perceived and to consider how we can train learners both to recognize and to generate visual representations of patterns of structure.

My collaborator, Jay Bolter, in his book *Writing Space* (1991) argues that these coherent patterns and links are elemental aspects of the associative nature of writing:

> No text is only a hierarchy of elements. A hierarchy is always an attempt to impose rigid order upon verbal ideas that are always prone to subvert that order. The principle of hierarchy in writing is always in conflict with the principle of association. One word echoes another, one sentence or paragraph recalls many others earlier in the text and looks forward to still others.... Associative relationships define alternative organizations that lie beneath the order of pages and chapters.... Previous technologies of writing, which could not easily accommodate such alternatives, tended to ignore them. The printed book has made the best effort to accommodate both hierarchy and association.... The table of contents, listing chapters and sometimes sections, reveals the hierarchy of a text, while the indices record associative lines of thought that permeate the text.... An index defines other books that could be constructed from the original book ... and so invites the reader to read the book in alternative ways. (22)

Bolter proposes that electronic writing, such as hypertext, an instance of topographic writing, is "both a visual and a verbal description... not the writing of a place, but rather writing of or with places, spatially realized topics" (25). Topographical writing is a spatial, visual medium as well as a verbal one. (Another reason I prefer the term *hypertext* to *hypermedia* is precisely because *hypertext* treats everything as topographical text; *hypertext* is the word's revenge on TV.)

"Although the computer is not necessary for topographical writing," Bolter notes, "it is only in the computer that the mode becomes a natural, and therefore also a conventional, way to write." For the computer "provides a writing surface with an extension and structure unlike previous technologies," one in which "topics... have both an intrinsic and extrinsic significance...they have a meaning that may be explained in words, and they have meaning as elements in a larger structure of verbal gestures" (16).

Frank Halasz (1987) focuses upon the distinctive interrelationship between scripted and discovered patterns of structure. Halasz notes that hypermedia systems, of course, must be able to search for specific content, i.e., words, key words, and so on. Noting, however, that "content search basically ignores the structure of a hypermedia network," he calls for the development of "structure searches" in hypermedia systems. As Halasz describes them, structure searches are not merely ways of seeking patterned coherence but are, in fact, ways to identify what Bolter calls elements in a larger structure of verbal gestures.

As an example of a "complicated structure query, involving an indefinite sequence of links," Halasz proposes a formulation that, though somewhat of a technical sounding riddle, is nonetheless quite visual: "a circular structure containing a card that is linked to itself via an unbroken sequence of 'supports' links." With a little study this verbal formulation discloses the structure of the riddle. "This query," says Halasz, "could be used, for example, to find circular arguments" (354). More important, as topographic writing in a hypertext, the visual riddle might likely be solved more easily than the verbal riddle of the query. In this case the authorial intention, or inattention, of a scriptor would disclose itself in a conscious search for patterned links, and scripted links would become discovered links.

As another of his seven issues, Halasz proposes the need for "virtual or dynamically-determined structures" (358) as a way to eliminate the "static nature of hypermedia networks." Again the verbal riddle is more forboding than the visual one here. Hypermedia networks are static because they only contain patterns and linkages that you, or someone, put there on purpose; they are what Halasz calls "extensionally defined" because the "exact identity of their components" are specified. For example, you might build a structure of everything your student Betsy wrote to you while you were away at a conference, but that structure isn't there until you decide to put it there. You cannot, in fact, see it.

A dynamically determined structure would just show up whenever you wanted it to. You would, in fact, have to query the hypertext at least once, building what Halasz calls an "intensionally defined" structure. To paraphrase Halasz's example, you would specify a sub-network containing all the nodes created by Betsy in the last week. Once that structure was created, however, every time you looked at it you would see what Betsy wrote during the last week. What you look for would be there to see when you looked. (This notion is not unlike what Alan Kay, the distinguished Apple fellow, has termed a "software agent.")

It is not too far-fetched, I think, to suggest that the ideas of Douglas, Balestri, Bolter, and Halasz open the possibility that hypertext learning tools may result in the discovery of what might, in a bad pun, be termed missing links, i.e., novel structures of thought and new rhetorical forms. What may seem far-fetched now is, I think, not just likely but certain. These new forms promise (or, if you will, threaten) to rival, or even supplant, the structures we have come to believe are more god-given than Gutenbergian.

This raises another peril, one perhaps more serious than over-promising. It *does* take some time to get used to, and to make use of, hypertext tools, especially when you use them constructively. Once you are used to them, they seem to adapt themselves so effortlessly to quite familiar, almost fleeting, and seemingly routine habits of mind that you are hard put to characterize, let alone schematize, them. Because of their fleeting quality, it is tempting to trivialize such habits of mind; because of the difficulty in schematizing them, it is tempting to presume them inferior to more established habits and structures.

To the extent that hypertext challenges traditional intellectual structures, it may be that this cardinal technology, like others before it, will threaten too much to unhinge us. We may perhaps, in the short term at least, lack the vision to appropriate these tools to the new tasks they suggest.

II

A true test of a challenge to traditional intellectual structures might be whether it is embraced as commonplace by those who do not feel themselves heirs to tradition. For the past few years my colleagues Cherry Conrad and Mark Harris and I have used Storyspace as a constructive hypertext tool with students who are often forgotten heirs of a passing tradition. Developmental readers and writers are students assessed as needing further work in writing and reading comprehension in order to succeed in college work. Unlike the technical or creative writing students with whom we have used Storyspace, these students have very little experience with computer programs, keyboarding, or writing environments and usually possess little or no conscious awareness of formal organizational structures for writing. Compared to these other students, developmental students make minimal use of the complex linking and on-screen hypertext capabilities of our program.

Nonetheless, I want to concentrate on my experience with developmental students, including the experiences of one student, in particular. I offer this account merely for what it is: not qualitative research or protocol analysis, not research or analysis by any means, but, rather, that most unstable currency, the teaching anecdote. By doing so, I hope to suggest that the challenge of hypertext to traditional structures can take on commonplace dimensions and that disenfranchised students, like expert learners, can use such tools to empower themselves in transforming knowledge to their own ends.

Teaching anecdotes of this kind tend to be small odysseys, accounts of where we went, what we saw, and, ultimately, how the world had changed upon our return. To have any sort of mythic power, however, these accounts require some measure of the gates through which we sailed and how we saw the world as we left. I admit to steering a course between cautionary and visionary pillars. It is cautionary and important in a realm of overpromise to keep in mind Guide inventor P. J. Brown's injunction that "those of us who expect the whole world to rewrite its documentation to fit the needs of our new hypertext system are unlikely to have our expectations fulfilled" (1987, 39). This is to say that, in using Storyspace with these students, as before when I had taught developmental writing, I set sail determined, again quoting Brown, to "fit the world as it is, rather than the world as we would like it to be" (39). Developmental students demand as much, although doing so in the most quiet and effective way possible, by ignoring, without indication or complaint, anything that their very practical experience tells them is bogus.

On the other hand, as teacher, writer, and software designer, I could not avoid navigating by the perhaps visionary truth of what Michael Heim calls "not at all extravagant" assumptions about the future: "Writing will increasingly be freed from the constraints of paper-print technology...and vast amounts of information...will be accessible immediately below the electronic surface of a piece of writing" (1987, 173).

What I saw in my student Les was an ability to see himself as freed of these constraints, constraints that he and the other students had, admittedly, freed themselves of long before either by rejecting them outright or simply by failing to learn. What had changed in Les and the others, however, was their ability to perceive and express, as easily as Heim does here, the existence of information below the surface of a writing and to use that awareness of structure in commonplace fashion to empower themselves.

This is obviously a long way to sail and a fantastical vision to claim to have seen. I should say how we got there. Like many others, the Developmental Education Department at our community college assesses entering students such as Les in the areas of writing, reading comprehension, and mathematics. The writing assessment is a holistically scored, timed-writing measure in which students write in response to a stimulus. Les's placement in English 101: Introduction to Writing, like most students, was based primarily on his difficulties with organizing a piece of writing and supporting his ideas in writing.

Students are assessed again upon completion of the developmental writing course, and, even before I began using Storyspace with them, my students already demonstrated a high degree of success in post-term assessments. I mention this both to avoid any suggestion of automaticity and to suggest another, obvious cautionary note—i.e., to the extent that the design of our hypertext writing tool was shaped by my understanding of the writing process, my students might be expected to succeed in using that tool.

Even so, to counterpose a visionary note, success, too, has its variations, few of which can be expected or explained as adaptive behavior. Les succeeded in unexpected and very simple ways, ones that I did not recognize at first and can only represent incompletely now. Quite unselfconsciously and routinely, this eighteen-year-old would-be auto mechanic grasped what was accessible under the surface of a piece of electronic (or, for that matter, traditional) writing and made it his own. Moreover, he did so in a way that I had not seen my students do before and that, in some sense, made me feel a generational alien, an outdated user of a tool I had helped make.

Mine is what might be called a neotraditional process model for teaching writing: i.e., like many writing teachers, I reject pre- and postwriting distinctions in favor of what is in my case a recursive three-stage model of form finding, focusing, and shaping. In the form-finding phase I encourage students to develop an awareness of "impulses toward form" (plans) and "transitory strategies for organization" (goals), using "interrupted automatic writings" to encourage a physiological awareness of shifts in intentionality (impulses or plans) and an ability to visualize and map them graphically on a handwritten page.

I likewise urge mapping of the consequent shifts in intentionality into provisional organizational structures (strategies or goals). During the first few weeks I use a progressively disclosed hierarchical sequence of abstraction—in which students are requested to consider objects, persons, and ideas in that order—to introduce the concept of "writing toward," an intentionally fuzzy scheme of organizational approximation.

My students are encouraged to map overlapping strategies (or goals) and to identify changes in impulses (or plans) at their intersections by drawing boxes, or frames on their writings. They are likewise encouraged, I should note, to map their own emerging understanding of the writing process against my language for this process, even the most jargony quasi-cognitive-scientific examples of it, which I use with them in the same way I do here.

On one level, the language is our shared joke against the world of the English teacher; on another, it is as mythic and empowering as any other part of this odyssey, a set of more-or-less understood names for skills they possess and have learned to recognize. In any event, this may explain why they are fairly undaunted when they come to use a somewhat complex computer writing environment—something most of my composition colleagues are curious, or even skeptical, about. Most developmental students have long suspected that something familiar lies behind the codes used against them, and they, in turn, have developed remarkably complex codes with which to defend themselves. Thus, they indulge me in my codes, especially when it comes to computers, which they view, rightly, as the ultimate code machines.

In the focusing phase, students use their cognitive mapping skills to help identify a network of interconnected, though not necessarily sequential, plans that lead to the development of the sequence of goals that will be presented to the reader, thus introducing a recursive notion of shaping. Students are encouraged to continue form finding recursively whenever sequences are not clear to them or might not be clear to an eventual reader, thus building a matching, recursive notion of form finding.

By the time Les came into my course I had begun introducing Storyspace to my students midway in the consideration of the second stage of the writing process. I showed Les and his classmates how to "port" the writings they had begun during the semester, as well as their mapping skills, to Storyspace. I encouraged them to use the program as a first stage of the shaping phase, developing an existing piece of writing for an audience—in this case in-class workshops. Then I more or less let them be.

By the time Storyspace came to Les it had gained some but not all of the features it would have upon its release early the next year. A set of tool icons is arrayed in the familiar Macintosh palette to the left of a scrollable, empty window in which the structure of a piece of writing and its links are eventually created. The first set of tools let you add and delete the fundamental building blocks of our program, what we call places, in the structure window. Places are individual, editable elements that can contain writing (including graphics); they may be combined into areas that indicate inclusion or subsumption. An area is a place that contains other places. The primary area is the document itself, which may be thought of as a nearly infinite space within which places may be created, connected, ordered, and referenced.

Other tools in this first set let you create automatically linked places for notes; select the pointer for navigation and other operations; use a powerful interactive mini-database to gather and link places in various documents; link places into paths across hierarchies in one or more documents; and choose among what we call "outline," "chart," or "map" views of an emerging document.

A second set of window management tools allows you to center places in a window; collapse, expand, or organize window displays; enlarge or reduce window sizes; restore the startup size window; and create a copy of the current window.

The third and most powerful set of tools lets you use a "block-buster" function to create new places from the text of any existing place; use a "block-maker" function to gather texts from the various places into one place; use a "place-marker" function to create a path of the places you work with or read; navigate through the displays and texts for your documents in either an editable or read-only fashion; get information about documents, places, or text; and link wide-spread places into paths across one or more documents.

Finally, various menu commands permit you to import and export texts from word processors or outline processors, distribute sections of text among places, generate new hierarchies from paths, and so on. Throughout its interface Storyspace operates on something of a principle of redundancy to counteract cognitive overhead. We have tried to design an inertia-less structure editor in which creating, linking, and rearranging parts of a document can be accomplished in a number of ways, according to your preference, and almost always visually.

Powerful word processors often allow you to adapt command keys and menu items to suit your preferences, but largely require you to work within a standardized environment. While Storyspace also allows a good deal of this sort of adaptation, it also lets you tailor the environment in which you work. Thus, e.g., the outline, chart, and map views of a document are much less specialized environments for different writing purposes than ways of graphically representing the provisional networks and transitory hierarchies that emerge and disappear as different purposes are discovered and developed. So too, multiple navigation schemes let you perform both content and structure searches of documents, searching for specific parts or exploring at will according to key words, place names, paths, etc. Searches are topographic, in Bolter's sense, according to both visual and verbal descriptions, and accessible through both visual and verbal browsing schemes.

Les, like most of our developmental students, worked initially in the map view, the most visual and topographic and the one most suited to creating constructive hypertexts. In the map view you create and name places as individual boxes, each of which contains editable text and graphics and each of which can be dragged and rearranged in clusters on the screen. This clustering ability, of course, offered an easy transition from the classroom work we had done. Clustering, to quote Michael Heim again, "reminds the writer and thinker of the sense of psychic wholeness in the world of increasingly fragmented texts and automated text manipulation" (1987, 176).

I do not wish to dispute Heim's contention that "clustering cannot properly be done in the computer interface," especially the kind of clustering he recounts "on a $57\frac{1}{2}$ inch piece of continuous paper," and especially since his view echoes the fervent criticism and battle cry of hypermedia theorist and IRIS director, Andries van Dam—i.e., "More screen real estate!" All this notwithstanding, our experience does suggest that the map view offers exactly the kind of "expandable graphic or map of thought discoveries" Heim describes clustering to be and further offers the "sense of wide open creative freedom combined with... peculiarities of connection" that he says "no software outliner could permit."

With all this as prologue, what Storyspace enabled Les to see may at first seem modest, if not illusory. This anecdote, when finally told, must be told backward, since I had little sense of what Les had been doing with Storyspace until he did it, and I have no memory of giving him any more instruction than the procedural help I gave others as they prepared their writing for "workshopping."

Les's first paper was an uncommonly good attempt at a fairly common writing, an autobiography in the form of an account of cars he had used, owned, been ticketed in, and, mostly, wrecked in traffic accidents. It began with his first "borrowing" of his father's pickup truck and continued through a "links," an Escort, another Lynx, and ended with his being given the pickup.

As we discussed the printed paper during the last weeks of the course, in a traditional classroom far from the computers, many of us liked one particular sentence toward the end of the paper: "Boy, if he would have had given me that truck when I had asked, this wouldn't of ever have happened." We all agreed the sentence managed to say a lot about maturity and desire. One particular paragraph, however, a description of one of these cars, troubled us. We were not certain which of two cars in succession it referred to, and we puzzled over suggestions for placing it, while Les listened in silence, according to the custom in my workshops. When it came his turn to respond to our suggestions, he addressed this one first.

"The box is in the wrong place," he said, "It will make sense when I move it."

None of us knew what to make of this, and a few students snickered quietly. Les was insistent and confident, moving on.

"I'm glad you liked the part about my father's pickup," he said. "It was the only arrow."

Neither the sentence in question nor the paper at large said anything about an arrow. The snickering increased, and I began to feel the kind of lotus-eater giddiness workshops sometimes bring as they veer out of control. Something in Les's insistence steadied me. He had lashed himself to a certainty we could not see.

"I'm sorry–," I said.

A rosy-fingered dawning came over him, and he grinned. "You don't know what I'm talking about, do you?" he asked. He scribbled quickly in the margins on his copy of the paper and held it up. He had drawn a series of boxes in the left margin.

"On the computer," he said, "that stuff about the car belongs next to the Lynx, and I moved the box but forgot to put it back. It makes sense in the right place."

What he was saying was that the structure of the writing existed electronically in a way that he could access and make clear to us, his assembled readers.

"And the arrow is a path," I said.

He was pleased that I finally understood and drew it too in the margin of the paper.

"Yeah," he said, "the only one. I noticed that it starts and ends with my father's truck, and so I put that sentence in."

What he was saying was that the verbal formulation of the paper led, topographically, from the visual representation of it. O brave new world. I had missed my first vision of it.

III

It may be that Conklin's implicit question, with which I began—what to do with hypertext tools—is the wrong question. Perhaps the better question is how to use these things to do better what you already do well. Certainly, the early proponents and true visionaries of hypertext believed as much. "The human mind operates... by association," claimed Vannevar Bush in describing his Memex in 1945, and, from Bush onward through Douglas Engelbart's Augment and even to Ted Nelson's Xanadu, the visionaries have insisted that the sometimes slippery and obscure trails of hypertext rest upon an underlying bedrock of natural cognition. With nothing less than democratic zeal, each of this trinity—the citizen-scientist, the engineer-rationalist, and the provocateur-humanist—builds upon a constitutional belief that habits of mind are naturally associative. They see hypertext trails as leading to a kind of shining electronic village upon a hill, an integrated, personalized, machine-enhanced, universally accessible, associative new yet familiar world, platted upon the patterns of synapses, deeded to each according to her or his needs.

It is a compelling, and potentially accurate, vision, and it is a vision I share. In the forty-three years since Bush's first exposition of this vision, it has attracted a litany of adherents. The IRIS project's Intermedia; Halasz, Moran, and Trigg's Xerox PARC NoteCards; P. J. Brown's Guide; Bill Atkinson's Hypercard; and, indeed, our Storyspace are each predicated upon equally democratic intellectual principles.

I wonder if we can hold onto them. Until recently the hypertext community—unlike the artificial intelligence community, for instance—has been able to make its case incrementally, without having to deliver upon strident claims and excessive hype. Ted Nelson's wonderful books and better talks aren't strident; rather, they're fervent (and, anyway, he baptized hypertext). My small odyssey and gossamer anecdote above borrow on this fervency, as does most anything written in this area.

Even Apple's marketing of Hypercard is less hype than an example of a conversion experience, a sort of corporate speaking in tongues—largely democratic, albeit accompanied by four-color tabloid testimony and Lucasfilm laying on of hands. Conversion experiences are common in the realm of hypertext.

I wonder what these conversions will cause us to overlook. Hypertext, unlike AI, has until recently enjoyed the safe harbor that relative obscurity brings. Yet education and technological change both stir winds across safe harbors, and now I wonder whether the claims we may find ourselves making about exploratory and constructive hypertexts alike may not put us prematurely in a whirlwind, not unlike the hurricane that accompanied the AI boomlet of recent years. Already the suggestion has been made to develop "a path analysis to classify prominent paths... learners take through a hypertext" and then perhaps to use "expert systems... to help learners access relative portions or sequences of hypertexts" (Jonassen 1988, 7).

Once I understood what Les was trying to tell me, I quickly secured his permission to let me investigate his workspace and study and report what I found there. While much of what he had done had disappeared in the process of his doing it, I found enough to think that we really ought to set up our wind machines, "instrumenting" and "journaling" the kinds of behavior he, in his own relative obscurity, had made commonplace.

It was interesting, for instance, to see that the autobiographical writing had sprung from a full-fledged writing environment. A first impulse appears at the end of a place called "Party Info" in a document he named "Family." The tale of his cars emerged from a typical account of a teenager's party, resulting from something of what Heim calls the "compensatory discipline" of "releasing" but which Donald Murray, Peter Elbow, Ken Macrorie and many other composition theorists have long had other names for.

It was even more interesting to note that he had at some point framed an organizational outline for the autobiography, deep within a map of another place, called "Christmas," which included places named for a sequence of gifts—"new toy," "new bike," "new car"—none of which he wrote about later. These uninhabited places upon a map of thought discoveries seemed to call into question the emerging body of research that suggests, as Christina Haas does, that "writers [plan] significantly less when they [use] word processing" and that "there was less conceptual planning and more sequential planning with word processing" (1988, 2). And they bring to mind Endel Tulving's speculations that "the kind of learning reflected in fragment completion and other similar tasks" might be "subserved by... an unknown [memory] system... the QM system (QM for question mark)" (1985, 3977).

Yet I wonder if all my and others' speculation will eventually help students like Les learn to do better what they already do well. And I cannot shake the uneasy, liberating feeling that the most dazzling and revolutionary exploratory hypertexts will be developed only when we come, like Les, to create constructive hypertexts that plumb the underlying topographical depths below a surface of boxes, one arrow, and who knows what other siren shapes of thought.

INTERSTITIAL

Networks of Woven Water—The Dangers of Transparency

Networks of Woven Water—A Minifesto for the TNC Polylog ([*sic*], i.e., what frogs sit on)

As part of a special issue of the journal *Writing on the Edge* devoted to hypertext fiction, Carolyn Guyer and Martha Petry consciously map and mediate intertextual relations in their collaborative interactive fiction, *Izme Pass*, which they describe as a "deconstruction of priority" (1991b, 82). "The purely arbitrary construction we decided to work within," they say, "is a triad of existing texts": two of them their own independent interactive fictions, "among which we began to weave a fourth, new work."

A genuine product of network culture, *Izme Pass* (Guyer and Petry 1991a) was composed by exchanging versions of their Storyspace document (itself an evolving network of parts of the triad of texts and interwoven additions), doing so over networks, all the while engaging in a constant collaborative conversation via e-mail and telephone. "What became most important to us," they write in introductory notes to the selection of e-mail exchanges that accompany the electronic fiction, "was how things are connected, not connection as conceptual negative space, but connection itself being a figure against the ground of writing" (82).

Electronic writing spaces open figure within ground in the way of reverse appliqué, illuminating what we see as we shape it. "The problem with getting inside the act of reading," says Jane Yellowlees Douglas, "is its ubiquity— there's no escaping it, and like any environment with which we are overly familiar, we no longer see it": "When we read print narratives we arrive already equipped with a full repertoire of reactions and strategies.... We never come face to face with the ground zero of reading. But we do reach that ground zero in [interactive] narratives" (1991, 120–21). The network always constitutes itself, even when its "subject" is expository or pragmatic, in the form of exactly such an interactive narrative. As Gail Hawisher and Cindy Selfe put it, "'Writing,' 'Reading,' 'Listening,' 'Learning,' 'Teaching'... in this late age of print seem blurred and overlapping as... technological innovations change what it means to be writers and readers, teachers and students" (1991, 3).

This blurring is in some sense the knowledge of the network, which, as Catherine Smith suggests of hypertext networks, "develops in a stream of successive, mutable fields permeated by influences from the organism and its environment." Thus, "a field description of knowledge," Smith continues, "would include the nature and dynamics of 'the inner life,' or *affective processes*: forgetting and denying as well as remembering and recognizing associations, rejecting as well as acknowledging connections" (1991, 239).

This field description of knowledge seems a conceptual rhyme with Petry and Guyer's deconstruction of priority, in which "connection itself [is] a figure against the ground of writing." Because the network as "a stream of successive, mutable fields" is always a deconstruction of priority, it discloses in its unfolding what we might otherwise take for granted on the printed page: the graphical juxtaposition and inversion of past and present that the tension (or weave) between print linearity and page design make possible, at what Patricia Sullivan calls "the point of visual-verbal integration, the balancing point for documents" (1991).

The network enacts this balancing point. "Reading ourselves... in a technological world awakens us to our roles, and our complicity, in the world," says Nancy Kaplan, playing off the formulation of Paulo Freire. "Our practical work must begin with reading the world," Kaplan continues, "but it must not end there, acquiescing to that apparently authoritative text in front of us": "Rather teachers must actively appropriate the world-text, and thus reinscribe—re-vision—the technology of the word" (1991, 38).

"I anticipate that our questions and answers will change in the asking," writes Martha Petry—"that we will have different utterances and tones depending on our emotional currents":

> I think of the writing as essentially flat, words on a page, icons on a screen, that become multidimensional as we layer, link, place,

> guard—our collaboration becoming a stream that cuts through the bedrock, a stream that is a continuously, synchronously, changing event where diving and stoneskipping, leaping across, and contemplating can happen simultaneously. We must listen so carefully to each other, how the stones splash. (1993, 83)

The Dangers of Transparency: A Note for a CCCC Think Tank Regarding Computer Intensive Instruction (San Diego, 1993)

The classroom is a mirror of a praxis and an expression of a *technē*, i.e., both a way to do things and an artful state of mind. How we know this is through learning and talk about learning, through teaching and talk about teaching, which is to say, through our recognition of the failure of institutions to meet human needs. Teaching so considered is itself a language like any other, performed in time, inscribed in space, mutable and entropic, subject to loss, misunderstanding, and failure to represent. Learning is such a language enacted and embodied; its meanings and uses are constructed in community.

The profession of writing has only lately (for two decades or so), and only uneasily, come to any agreement on what the classroom mirrors or if it mirrors at all. As is true of the uneasy agreement we form about any language, much of the agreement about what teaching does has been driven by our ability to recognize its failings. If we construct meanings and derive uses for learning within a community, we look where the transitory community of the classroom fails to know where learning fails. Whether in failing to represent diversity of experience, varieties of cognition, emotional complexities, power relationships and interdependencies, or the terrifying and heartening inability of our species to signify except through multiplicity, we know that it is learning that fails, not learners.

In recent years many of us have come to value the ability of computer learning environments—taken in the widest sense, from software, to computer classroom, to synchronous conference, to internet, to hyper- or cyberspace and virtual reality—in representing learning. Writ upon the technology of the spatialized classroom, failure shows itself like dust on a mirror. A voice that does not find a fit in a classroom editing session, a computer conference, an e-mail exchange, or a hypertext collaboration shows through, stands out, takes place, transforms, reconfigures. Again, it is learning that fails, not the learner.

More important, with careful attention to where the voice that fails to fit places itself, the mirror metaphor fails and gives way to one of intervening, aqueous lenses whose translucence shifts from moment-to-moment between glass and mirror. We are able to sense how many nested, transparent surfaces construct the apparently singular mirror. We are able to map the multiple and interstitial contours of discourse, of power, of desire and necessity, that the classroom is called upon to serve. Like the lensed eye, the classroom reforms itself to focus upon what previously could not find a fit.

Sometimes this reformation is a happy one, a renewed vision in which we see through disenfranchised, marginalized, or simply more quiet eyes. Other times it is an uncomfortable, though no less happy, reformation, in which we recognize what clouds or obscures our learning. Often what cannot find a fit in computer classrooms is teaching itself or else the needless technological demands of classroom systems.

Networking software, which requires that classrooms be represented as hierarchies of ownership and privilege—folders within folders—fails to account for the increasingly unbounded decenteredness of computer learning. Conferencing software, which won't let us pluck a comment from the scrolling stream of synchronous postings in order to see its fins and feel its flank, does not account for our ability to balance the energy and immediacy of interaction with thoughtful contemplation. Hypertext software, which forces us to represent contours of interaction in rectangles and arrows plotted on Cartesian space, fails to account for the gleeful vertigo we feel on first coming to see cyberspace and our ensuing desire to shape even discursive space proprioceptively and sensually, i.e., as the body knows.

Similarly, the disjunctions, imbalances, and resistance involved in instruction are much less transparent than they were in classrooms without computers. Teachers who wish to channel, shape, amend, correct, challenge, or edit what streams by in synchronous conferences take their chances in discursive time and inscribe their power upon the (often disproportionate) space their comments take up before they disappear into the upward scrolling windows.

Likewise, teachers who would comment on a hypertextual link must often, instead, interrupt it with a node of commentary that deflects or defuses what they meant only to herald or mark, while artfully designed visual webs of hypertextual reading, collaboration, and commentary require that teachers literally find a position within which to write and to understand that an equally artful placement earns them only equal standing in a dimensional map.

Finally, teachers in an internet polylogue find that their students as well as other learners and teachers are paradoxically both connected and separated, both localized and undifferentiated, by virtual time and space. To claim that one's teaching is decentered in such space seems only to brag and beg recognition of what authority has, in fact, been passed on. To discharge institutional responsibility to one's students means intervening in (and even helplessly subverting) the newly emerging transcendent institution that the network colloquy enacts and literally represents.

Technology aspires toward transparency. Insofar as that aspiration intends to hide its failings, technology, like any unacknowledged representation of power, endangers learning. This is not to say that we should be cynics or Luddites—that we should expect learning failures or that we should oppose technological progress—only that, within the crystal palace of an increasingly transparent technology, teachers face increasingly paradoxical tasks. Like Dorothy parting the curtains of Oz or the Halloween goblin soaping windows, teachers may seem to stand in the way of their own progress or work against their own interests on the sweet road toward home. To each new task that technology accomplishes automatically, a spatializing pedagogy should append a step that maps its undoing, i.e., what is no longer seen and whom it affects.

Such a step seems obvious and even necessary in helping a beginning writer understand that a spell-checker is a kind of video game wherein alien algorithms sometimes infect healthy colonies of living words. There we gleefully map a missed homonym that was wrongly given a "waver" or *unparsed plurals,* which the program changes to *unpaired pleurisy.* Yet consider when Apple, e.g., in response to "a clear demand... [from] classroom users," announced a "workgroup server" complete with built-in (proprietary) " 'agent' software... that can be trained to perform some specialized task," such as "to search through thousands of documents quickly and find the closest relevant match to a particular query" (*New York Times,* 22 March 1993, D1).

In the face of all this transparent and welcome utility it will take a tiresomely stubborn teacher, and a desperately complex parallel sentence structure, to recall to her students that learning is a language enacted and embodied and that the meanings and uses of words like *clear demand, specialized task, closest relevant match, particular query,* and *dangers of transparency* are constructed in community rather than ROM. Even then, who will hear her, through the muffling thickness of mesmerizing glass, isn't clear.

What the Fish Lady Saw: Patterns Out of Disjointedness in Two Hypertext Writing Communities

the stuff about hypertexts and moons and computers is that sometimes they have their own way of leading us where we are not sure where we're going.... surprises, pits, interesting conversations and images—hypertexts... that sometimes make us listen to what we least know

—"*fish lady*"

It is sometimes possible to see an idea actually come into being. At the 1988 Computers and Writing conference in Austin, I sat across the aisle from Gail Hawisher during the Q&A following the landmark standing-room-only session about hypertext presented by John McDaid, Stuart Moulthrop, and Terry Harpold. Even now I can recall the mix of perturbation and delight in Hawisher's expression as she struggled to form a recognition into a question. "I am beginning to see," she said as my memory now paraphrases it, "that collaborative writing and hypertext and computer conferencing are somehow converging into one process" (see Hawisher and Selfe 1991b).

Within the convergence of electronic classrooms and writing environments, expert-novice distinctions between groups of writer-learners (including the expert-novice distinction of teacher-learner) seem less useful than do characterizations of the richness, multiplicity, and power relationships of the writing communities that they form (Cooper and Selfe 1990; Hawisher and Selfe 1991a; Moulthrop 1991). Indeed, it can be argued that the decenteredness and simultaneity of electronic classrooms recast teaching as continuous discernment of the patterns of community that form around writing tasks and shifting views of the roles of reader, writer, and text (Johnson-Eilola 1991; Joyce 1992; McDaid 1991). Not only is it sometimes possible to see an idea actually come into being; it may also be that, in teaching in the posttechnological age, the idea of teaching is finally transmuted into discernment of the learning process.

Thus, although I hope it will shed a good deal of light on such questions, this research is not focused upon how well or even how—or, for that matter, if—students learn in networked collaborative hypermedia classrooms. Nor by any means is the comparative data of this study meant to suggest that one or the other of the two writing communities studied is in any sense superior or deficient in its interactions or inquiry. Rather, the two classroom communities are counterpoised as arenas of discernment. Likewise, though it will certainly shed much light on my own attempts to discern, this research does not focus upon discernment per se. It is, instead, an attempt to discover the degree to which two hypertext tools—Storyspace and Eastgate Systems' experimental tool, Link Apprentice—offer appropriate means for discovering, analyzing, and mapping dynamic learning processes undertaken in collaborative writing environments using a third tool, the Interchange computer conferencing (ENFI) feature of The Daedalus Group's Discourse program. A research agenda for this kind of inquiry has been spelled out in detail by Nancy Kaplan (1991, 36–37):

- What statements do[es] a . . . complete "writing environment". . . make about the nature, function, and power of the writer, the text, the reader? And how do these messages "fit" with the other ideological strains of contemporary writing instruction?
- . . . What is the relationship between the conceptual fields a particular configuration of hardware and software opens and those it obscures or proscribes? . . . What privileges and prohibitions are embedded in the design of the particular tools used in the studies? How are writers' behaviors, and even their desires, affected by the workings of the tools' deepest structures?
- How do alternative modes of discourse—the new possibilities of hypertext and hypermedia structures—work, rhetorically and ideologically? Whose interests and visions, whose realities, do these structures serve and whose can they be made to serve?

And, finally,

- What part does the emerging networking environment play in the process of forming and validating discourse communities?

> Are hardware and software configurations working toward the maintenance and the perpetuation of existing hierarchies of privilege . . . ? And what alternative can teachers imagine and create?

Population. Writers in two Jackson Community College (JCC) classes were invited to consider their experiences in networked, collaborative, hypertext writing environments by participating in synchronous conferences with other members of their learning community during the fall term of 1991. Each class met separately to discuss their experiences in writing within electronic classrooms. A first-term composition class (hereafter called Comp Writers) considered their experiences midway in their first term in beginning to write with Storyspace, while a Creative Writing class (hereafter HT Writers) did likewise midway in their third or fourth term, writing, analyzing, and helping others write electronic hypertexts and conventional printed documents using both Storyspace and Discourse.

For the discussions analyzed here students in the composition class met during a one-and-one-half-hour scheduled class session, while the hypertext writers met outside class hours in a late-afternoon session that lasted approximately three hours (with only four participants, themselves Interchange habitués, remaining for the last hour). Each class was aware that the Interchange sessions were intended for this research, and each class member had signed a human subjects research agreement. As always in Interchange sessions, the students were able to assume and switch at will among pseudonyms of their choosing. Although I participated in these sessions, I did not use a pseudonym, other than my usual, known Interchange "handle" of Mickey (the only "place" on earth I have ever used this name).

Since the participants used pseudonyms, it is impossible to give an accurate account of participation and involvement. Nonetheless, median number of interchanges give some sense of the difference in participation levels. For the 13 HT Writers, who largely stayed with their usual handles except when changing names for rhetorical effect or fun, the median number of interchanges was 22.0 (N=367). For the 36 names of Comp Writers listed in the transcripts (from a class of 25 members), the median number of interchanges was 4.5 (N=271). The median for Comp Writer names participating more than two times (and excluding my and the classroom tutor's interchanges) was 8.5. Finally, for the three "active" writers in each group singled out for link plot analysis below, the median number of interchanges for the HT Writers was 41.0, for the Comp Writers 12.0. No data was collected regarding the sex of participants, although one's choice of a male or female handle is a fairly reliable indicator; nor was age data collected. Both classes were young by community college and Jackson Community College standards (the average age of JCC students is 29.5), although the Comp Writers included four students over thirty years of age, one of them a woman in her fifties. Finally, it is important to note that the Comp Writers had no indication before their first class meeting that the class was scheduled in an electronic classroom and that, at the time of this research session, none had dropped or stopped attending the class.

Methodology. Transcripts of the two Interchange conferences were organized into a Storyspace hypertext in order to characterize the writing communities and the issues that engage them (Joyce 1992b). The organization process involved importing transcripts of the two Interchange sessions in Storyspace and using the program (1) to create a separate space (box) for each interchange; then (2) link all the interchanges together in their linear order; (3) gather spaces together in groups corresponding to my perception of emerging themes; and (4) assign key words that identified names and listed analytical terms for each interchange; while also (5) writing descriptive summaries of the interchange discussions; and (6) methodological notes regarding my protocols and process of discernment. The descriptive summaries formed the basis for extended ethnographic narratives that are excerpted in the following pages. These descriptive summaries serve as both a qualitative report of the discernment process (see Douglas 1991, for this method) and as a basis for evaluation of the Storyspace and Link Apprentice tools for visual analysis and discernment, which also appear below. Finally, using the lists of key words spawned for each group's transcript, the Storyspace "Path Builder" was used to generate paths through each document corresponding to recurring key words and combinations of themes within the discussion that I deemed important. Paths (or trails, links, webs, etc.) are the essence of hypertext, which I have elsewhere defined as reading and writing in an order you choose and in which your choices constitute the current state of the text (Joyce 1991c; see also Bolter 1991; Landow 1992; Nelson 1987, 1990; and Neuwirth and Kaufer 1989). The Link Apprentice tool used these paths (created by linking key word–coded sections of the class discussions) to generate "linkplots" that offer a graphical representation of the complexities of the respective discussions.

Key word coding. Key words were assigned during the analysis of the interchanges, sometimes by using a Storyspace tool to add a word used by the writer, in an interchange, to the key word list, other times by using the program's key word editor to assign existing key words from an existing list or to assign new key words not already in the list or in the text of the interchange. Obviously, as the list of key words increased, the likelihood that an interchange would suggest a new key word decreased. So too, the likelihood that later interchanges would be assigned several existing keys was increased (although I also regularly used the program's text search and navigation capabilities to go "back" into the hypertext transcript and assign more keys to earlier interchanges). Since the focus of this study is upon the process of discernment, I developed no standardized set of, or taxonomic scheme for, key words before analyzing the hypertext transcripts. Nor did I make use of the excellent evaluative schemes developed as part of the ENFI project (Batson 1988, 1989).

As a methodological choice, however, I did analyze the HT Writers hypertext transcript first on the premise that their discourse would likely provide a rich set of keys to use in analyzing the Comp Writers. Also I adopted a few standard key words early on as shorthand. For instance *affirmDis*, the most frequently used key word, was assigned for any interchange in which the writer consciously acknowledged the other writers or the process. As my analysis uncovered rhetorical features common to the interchanges, these too were assigned "standard" key words (e.g., *agenda-setting*, was assigned to those interchanges in which, in my judgment, the writer attempted actively to focus discussion in the Interchange, whether on a serious or less serious subject; other keys, such as *paired-comment* and *query*, emerged as useful descriptions for regularly occurring kinds of serial interchanges adopted for a given effect and rhetorical purpose, respectively). As noted, key words form the basis for the spawned Storyspace paths and subsequent Link Apprentice linkplots, yet even a rough comparison of the frequency of selected key words offers some immediate ground for distinguishing the two groups.

The Comp Writers session, e.g., was more focused upon the collaborative writing process (*collaboration, discourse, spelling/typing,* etc.), while the HT Writers session centered on technocultural issues (*"the book," interaction, tradition, computer, etc.*). Comp Writers more frequently acknowledged the interchange process (*affirmDis* 78.97 percent of total spaces) than did HT Writers (65.12 percent) and were equally likely to focus discussion (*agenda-setting*; Comp Writers 19.93 percent, HT Writers 21.53 percent). The ratio of reading to writing keys within the interchange sessions seems especially telling—with HT Writers much more likely to consider the reader (Comp Writers 0.26, HT Writers 1.50); and is, I think, supported by the following descriptive summaries.

Descriptive summaries and methodological notes. As noted, I analyzed the HT Writers' hypertext transcript first on the premise that their session would likely provide a rich set of key words to use in analyzing the Comp Writers' session. By first gathering and grouping spaces corresponding to my perception of emerging themes and then contemporaneously writing descriptive summaries of the sessions, I hoped to develop a similar strategy. This did not prove to be the case. Grouping and summarizing the HT Writers' session proceeded with very little difficulty—not surprisingly, since they not only shared my intellectual and artistic interests but also had worked with me for a number of terms.

The descriptive summaries became highly distilled redactions. I retained the spellings and punctuation of the transcripts in quoted interchanges, partly out of Elizabethan fervor (Joyce 1978) but also because it is the custom of electronic communication, even academic communication, that surface structure is not a matter of consideration (another electronic custom I've maintained here is more evaluative, using *says* rather than *writes* to describe a contribution to an interchange). Some of the orthographic diversity in the summaries is an artifact of the Discourse software, which in large groups slows typing and for the unwary participant leads to collage of text fragments—and to complaints in both sessions.

The only real methodological decision was what depth of the structure to maintain for the analysis, since these writers were likely to embed (or umbrella) theme within (under) theme, in the process known as "threading," among participants in electronic conferences and discussions. (I use these terms interchangeably, no pun intended, with *thread* more likely to refer to the actual comments regarding a discerned theme.) Since it kept the HT Writers' summary compatible with that of the Comp Writers, I maintained only one level of organization. For this and other reasons (e.g., a wider perspective for discussion themes) twice as many thematic spaces were created for the HT Writers. Thematic spaces appear in the following excerpts in the following form, {*beginnings (8)*}, in which my title for the thematic space, "beginnings," appears first followed by the number of interchanges in the section (8). The following excerpts from the HT Writers' summary give some flavor of the complexity of their discussions (one space, ***{choices again (31)}***, is boldface to indicate that it is the subject of further linkplot analysis at the end of this chapter; full transcripts are included in the appendix).

> From the Descriptive Summary—HT Writers. {*Agenda-setting (9)*}. Housekeeping and hellos. The group affirms its customs, makes certain that a particular discourse won't be constrained. {*Definite form (7)*} Bill attempts to set the topic to discussion of form, but the flow of introductions, delayed agenda-setting, etc., deflects him. He restates and slightly alters his theme. {*Resetting agenda (8)*} A short exchange asserts the accustomed patterns, i.e., "say[ing] things that are totally irrelevant," against Bill's foray, which he repeats. {*Definite form 2 (7)*} Dawnkiller picks up on Bill's theme of form and offers a definition of hypertext, "emotion . . . seems to be the embodiment of hypertext." Zøʃó attempts to pick up my strand about visual form, making a distinction between hypertext reading and writing. The others parody these discussions with a continued thread about flies (and pies?) in which Boomer connects the perception/interruption of a fly buzz with hypertext reading as Dawnkiller raises the issue of real vs. virtual books. {*The "book" (8)*} Moon picks up on a note of Dawnkiller's referring to a past Interchange session regarding hypertext and writing, the transcript of which had been lost. (Moon's report of that previous exchange in some sense precipitated this session.) Now she proposes to raise the issue of books vs. hypertext again. Others, however, seem more inclined to fol-

low on either Bill's discussion of form or my question about visual writing. Dawnkiller merges the discussions in a brilliant gambit regarding internal form and the nature of the book. Meanwhile, Bill's and ZøJó's talk about beginnings of hypertexts will both provide the next theme and merge discussion.

[transcript sections deleted]

{choices again (31)} The weave continues (crescendoes) with a four level discussion of (1) how to disclose choices to a reader and whether the reader can actually take control, (2) what will be the future of the book, (3) why the college library burned books removed from the collection, and (4) what is different about reading and writing hypertext. As a result (or so I hypothesize), I collect more spaces into the section of the analysis. Perhaps, however, it is only that I am already tiring with the more granular analysis, only a third of the way through the transcript. In any case Stephanie finds a way to enter the discussion in this section (and each time is mentored by Bill). Moon is seemingly (naturally?) unaware that the discussion continues under the "whither books" umbrella she has raised much earlier, and so she restates it formally here midway, again referring to the lost, past Interchange session. Wabbit and Dawntreader tie teaching, hypertext, and the future of the book together in a thread about knowledge, partly in response to Stephanie.

[transcript sections deleted]

{hypertext critic (47)} The consideration of Lestat's "opposition" takes something of the form of a Platonic dialogue, sharpening consideration of its subject and in the process shifting its meaning subtly and unalterably. Lestat's resistance to reading on screen gives way to his argument that hypertexts lack quality in their writing, especially as measured against "real books." Moon and Boomer question whether Lestat has ever written in hypertext, and he admits to being somewhat curious about it, especially as it functions in an "annotative" way or like a "text video game." They wonder together what changes in reading technology will come in the future and whether hypertext is merely a "fad." Moon assumes a role of defending hypertext as a writer's medium but not as a reader's (both because it is not portable and because she does not want to see the book replaced). This puts Lestat (who, it seems, in the previous spring's discourse took the "adversarial" role that book burning is okay because there are too many books) in the position of defending hypertexts as less wasteful delivery vehicles for technical writing and education. Moon agrees about "tech" writing but still is loath to see the book go. An interwoven strand throughout this section reconsiders the question of the writer's intentions. Early on Lestat argues that "an author has a certain direction s/he wants to move the reader in." Uther and Lestat thereafter conduct an interwoven dialogue about reading order and how the reader forms the text, which culminates in what the fish lady said.

By the time coding and key wording of the Comp Writers' session was complete (but before spawning paths from key words and linkplots from paths), I grew aware that the fine strands of the eventual threaded Comp Writers' themes appeared early in the interchanges. Yet on first (and sometimes second and third) pass I did not see this and, in fact, made an early decision to gather and title thematic spaces at points at which it seemed to me that one student or another had actively engaged in agenda setting and to put off writing descriptive summaries until the gathering was complete. I.e., I believed I was bookending fairly disparate sequences between agenda-setting comments. Even so, before I was done I found myself consolidating the sections even further, discovering, e.g., that the last six thematic spaces were really three spread out over more distance. This methodology was improvisational and quite the opposite of the methodology I adopted with the established community of HT Writers. With the hypertext writers I wrote these section summaries after analyzing each section, largely because the interchanges seemed complexly interwoven and necessary to summarize while they were fresh but also because the coding itself led me to a process of discernment. (Using the Storyspace anchored windows feature, I wrote summary notes as I cycled forward and back through the gathered interchanges or voyaged, via search mechanisms, to earlier thematic spaces.)

The Comp Writers' interchanges seemed both more dispersed and, paradoxically, also more sequenced and discrete. The linkplots and the descriptive summary (excerpted here) alike affirm the paradoxical nature of the Comp Writers' discussions:

From the Descriptive Summary—Comp Writers. {*preliminaries (12)*} After initial Interchange sessions early in the term this class has largely done its group work in Storyspace. Thus, there is a great deal of recalling how to get started. Django's "This hypertext is very hard to understand" seems the kind of formal foray by which, even at midterm in a supposedly decentered, self-grading, collaboratively grouped class, students test your intentions and the reality of your conversion: Django knows I need research; he poses a research question. Given a ground, I build on it, framing my interests in two parts: "how this kind of writing environment changes your view of the writing process" and "what... you think hypertext is (if you think anything at all) [and how]... do you think about the parts of a writing... in a Storyspace?" The Stud and shazbat begin the thread I've key-worded as *penileHumor.* {*agenda hunt (22)*} Given the piles of fleece laid out by my agenda setting (and the dark fur of the Stud's anti-agenda), a number of participants begin to hunt for a thread that will engage the group. Emotionally Ignorant considers collaboration: "Determining whether or not this works for you depends on the group you are in and if they work well together." Denzel suggests that "the computer has helped my writing... [it] makes me aware of spelling, and form." Squirrel merges formal advantages and group advantages in a happy formulation: "It helps cuz there is more than one persons input on a paper instead of just one, so no ones weekneses realy show up in the document." No one spins at the hypertext thread directly except the agenda-setter Django, who, despite his early weighing in regarding its difficulty, now "want[s] to know what hypertext is." Air Man—in an interchange that at first orthographic glance would seem to confirm the suspicions of the deepest skeptic—eloquently evokes the spirit of hypertext and collaborative work in a fine insight about writing processes: "The exptation of this type of a writing inverment is to be able to write whatever you think when you think it and arrange it when you nead to." When I realize now that each of these threads—formal advantages, group success, and symbolic manipulation (i.e., writing/arranging chunked, though)—remains a matter of active inquiry through this whole interchange session, I am surprised, unaware either in the session itself or in my coding pass through this material, of these thematic threads.

[transcript sections deleted]

{*re hypertext (30)*} The group ranges through a halting attempt to discuss my question about whether a hypertext writing environment really lets you choose what and when you want to read and write. Z joins in and (I think now) models typical Interchange style by narrating his experience with collaborative hypertext. I restate the concern as "What is different about writing with Storyspace? what is the same?" Since it follows only two interchanges later, Zoe's stunning response has likely been something she was composing through the series of jokes that followed my initial agenda setting. Her response is an example of what I've key-worded as "set pieces" and, in fact, a lovely consideration of the question. Only now really, as I write this summary, do I discover this and add *hypertext* to the key words for this response. Her dream meditation will resurface late in this session in a thread about reality:

> Zoe: When I lay down at night my mind slips back a few feet wondering about the day I just experienced and then jumps forward about ten miles. I can't keep up with it, sometimes I think I have insomina. I don't care if I ever sleep again as long as I have intense dreams, I could spend the rest of my life dreaming and wondering if I spelled *insomnia* correctly.

Air Man also continues to advance this question in his (orthographically more kosher) observation that "for a mixed-up thinker who thinks randomly this is the way to write." Denzel likewise expands on his initial insight that escaping surface structure concerns frees one to consider real (rhetorical) structure. "When I write on paper it seems to be alot of jumble," Denzel says, "I not only have to deal with puncuation but with my own handwriting. But with the computer I have less to deal with so the structure has a better foundation. And I find that people can read my writing with greater understanding." Finally, the aptly pseudonymed Aristotle, in an echo of Hawisher's question, asks "Is this (discourse) part of hyperspace or Storyspace?"

Although I took a much more active part in agenda setting and questioning in the Comp Writers' session (40 interchanges, 14.8 percent) than in the HT Writers (18 interchanges, 4.9 percent), when I began gathering Comp Writers' interchanges following an establishing agenda-setting comment, only the first agenda-setting comment was mine. I am not certain whether this is because others actually took over agenda-setting as the interchange went on or whether I was more able to see a way to organize my hypertextual summary around their concerns. I am aware that I also set the agenda for the HT Writers' session, but their treatment of issues I raised was somehow different. The hypertext writers washed over my agenda-setting in successive waves of discourse, while the Comp Writers checked off items in the query/response way. The Comp Writers regularly issue such queries and await an answer in the interchanges much in the same way one would search a database. The HT Writers almost never do this, and, in fact, I didn't code a key word for them corresponding to the Comp Writers' *query*.

Another example of differences between the two groups in interchange rhetoric is the extended writing that I've key-worded as *set pieces*, something that the Comp Writers do and the HT writers almost never do. The most spectacular example is sassi's set piece, which sets the agenda for the *changing writing* section of the Comp Writers' discussion.

> {*changing writing (31)*} Of the many remarkable things about sassi's set piece, perhaps most remarkable is how little immediate effect it has upon the discourse (although it becomes a deep flowing stream that shapes following interchanges). Nonetheless, the set piece is a wonderful example of a fully formed improvisational composition. Beginning with a paragraph-long consideration of technology itself, how learning "that I don't have to be affraid of the computer... I can us it to do things that I would have never found out about [and]... to get intouch with the new world." She then moves to another paragraph-long consideration of collaboration, which she values because "there are so many interesting people in here... [who] do really care if you succeed or if you just need their help they are there for you...." Next she reframes Denzel's earlier point about how escaping surface structure concerns frees one to consider real structure, noting that

> "writing on computers... you can move around paragraphs so that they might be more interesting to read. It is not as stresful as when you longhand write and then you find that you want to rearrange your paper and that means that you have to rewrite your paper all over again." After a bow to the competitive advantage computers might give in a technological age, sassi closes with a query: "I'm glad that I took this class.... What about the rest of you???" Even her rhetorical question (unlikely to pass traditional classroom scrutiny) can be seen as beyond the essay, a punctuating query that announces sassi's return to the flow. Since the flow is constant, extended replies like hers get less attention than shorter screens do in electronic writings (Interchange, Storyspace, or e-mail). The group has unwittingly gone on to consider the issues sassi raises (the "mixed-up" thought of hypertext, e.g., or Denzel's complaint that he "get[s] typing some times fast and mistype words"), but sassi must make her points in the flow again as she did in the essay. She does so ably, telling Denzel, "Don't worry about misspelling words and putting things into improper sentence form that you can fix anytime," and responding philosophically to the hypertext thread: "Life goes on any way you can get it to go and then some...."

The occurrence of these set pieces may suggest that the Comp Writers are more likely to look on electronic writing as presentational and formal than are the HT Writers, who see it as conversational and "intertwingled" (Nelson 1987). Or perhaps the Comp Writers interact by gathering ideas that they broadcast in set piece quanta, as opposed to the more wavelike interaction of HT Writers.

Linkplots and maps. The balance point between access and control in converging technologies is dynamic and, so, difficult to fix. Learning processes like those discerned within these descriptive summaries require ephemeral, easily mutable structures to represent them without quashing their autonomy. Otherwise, as Kaplan suggests, the structuring itself maintains and perpetuates "existing hierarchies of privilege." Synchronous conferencing approximates a flattened hypertext, while hypertext in turn spatializes (literally re-places) temporal structure. Both indeed seem to suggest the kinds of mutable structure necessary for representing learning and transmuting teaching into discernment of the learning process. Yet it should also be possible to gauge the adequacy and reliability of the learning process *in process*—i.e., in Julia Kristeva's method of toccata and fugue, "sketching out its perpetual motion through some of its variegated aspects spread out before our eyes" (Kristeva 1990, 3). We need to be able to fix the balance in the way of the descriptive summaries, but doing so improvisationally, on the fly, when access to the full transcript is either not possible or not desirable.

Both the topographical writing space of Storyspace and the linkplots of Link Apprentice are attempts to gesture toward certain features of a wealth of structure—complex, incomplete, elusive, but unmistakable—that we can only partially discern and only then at great cost in investment of our time and analytic energies. The figures below reveal patterns of interaction as expressed by paths through each document created in Storyspace and corresponding to recurring key words and combinations of themes deemed important in the discussion.

The linkplot in fig. 1 shows each hypertext link as a dot. Every writing space is assigned a number in sequence (in this case, the number is based on the position of the writing space in the gathered hypertext transcripts), and a link between, say, writing space #47 and #63 is represented as a dot at 47,63. The broken central diagonal indicates that the hypertext transcript was arranged in a sequence following the time line of the sessions. Unless all links are bidirectional, the linkplot will not be symmetric; points above the diagonal represent analytic links (and, indeed, thematic continuities) that jump "forward" in the re-placed temporal structure of hypertext transcript, while links

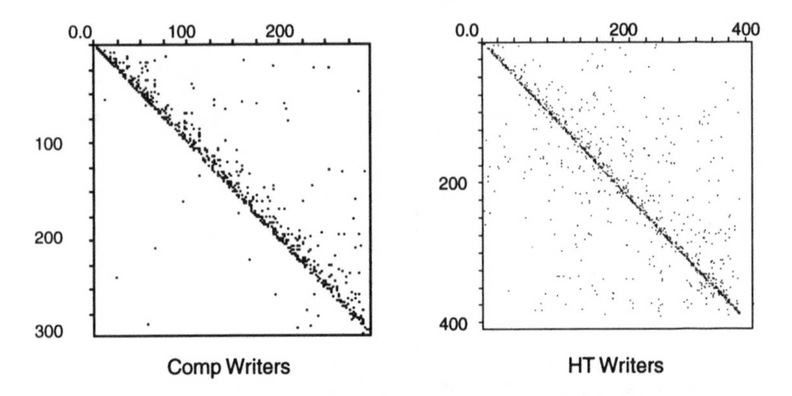

Fig. 1. Linkplots of the full sessions; each dot represents a single link

below the diagonal represent links that jump "backward" with respect to the temporal ordering that underlies the transcript.

Thus, the regular occurrence of links along the diagonal offers a baseline (time line) for each discussion. If we imagined the extremely unlikely possibility of a discussion in which each coded section referred (linked) only to itself, the resulting link plot would be a single diagonal line of dots. If we imagined the equally unlikely possibility of a discussion in which each coded section referred (linked) to every other one, the resulting linkplot would be a solid black square of dots with no discernible diagonal time line. Thus, the complexity of a linkplot suggests the degree of linking discerned among widespread comments in the discussion. Such a diagram is the graph-theoretic adjacency matrix; raising this matrix to, say, the fifth power yields a plot that reveals places separated by no more than five links in the gathered hypertext transcripts (fig. 2) and has the effect of visually enhancing the degree of linking discerned.

Dark areas in the linkplot identify clusters of writing spaces that are closely associated, while light areas highlight spaces that are sparsely interlinked. Clusters of dots in figures 1 and 2 clearly suggest the thematic divisions of the two sessions. In the Comp Writers' sessions (especially in fig. 1) the nine thematic sections (or threads) discerned in the key-worded hypertext analysis are visible as something like sunspot blips above the diagonal horizon line. In figure 2 these threads merge into a smear along the horizon of the diagonal, which

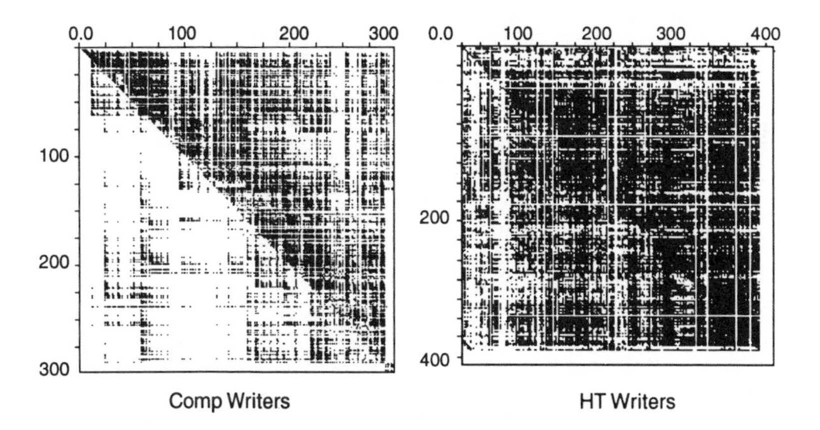

Fig. 2. Augmented linkplot of the session; each dot represents places separated by no more than five links

nonetheless suggests three or four large organizational clusters along the time line. In the HT Writers' sessions the densely packed intertwingledness discerned in the key-worded hypertext analysis is apparent even in figure 1 and is resoundingly clear in figure 2, which resembles a voice print of the Tower of Babel.

*Linkplots and maps {**choices again (31)**}.* Linkplots attempt to represent ephemeral, easily mutable, large-scale structures in a patterned, gestural fashion. So, too, Storyspace maps trace the succession of advances, accumulation, construction, and replacement that form the contours of emerging structure. As an example, the multilevel discussion regarding readers, choices, control, the future, the book, writing, and hypertext that engages HT Writers in the *choices again* strand appears—once it has been thematically linked—as something of a whirlwind in a reduced Storyspace view (fig. 3).

This whirlwind can take up the reader at many points, and it can traverse the ground in many ways. The Storyspace view maps contours of thought whose vectors appear as a convergence of arrows linking voices. At normal size the map reveals that the contours continually converge upon *moon, bill,* and *dawnkiller,* three primary voices among the nine who participate in this strand.

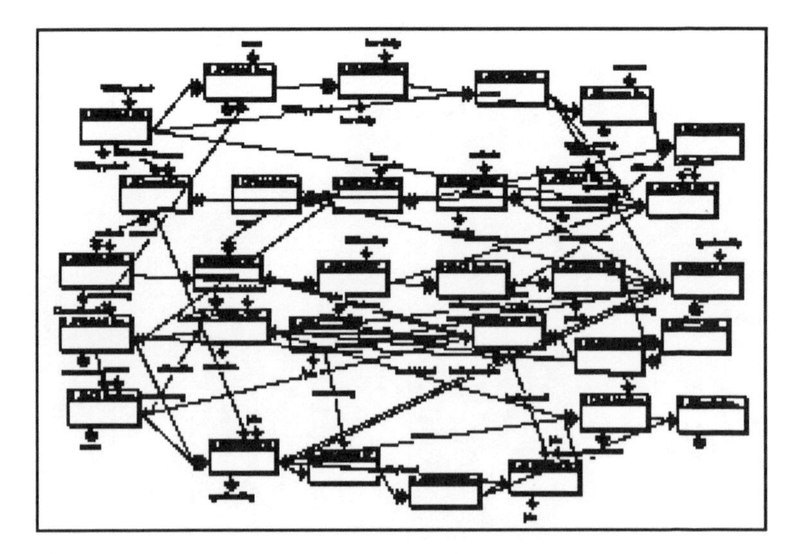

Fig. 3. Reduced Storyspace view of HT Writers' thread {*choices again*}; each square is a mutable text of one interchange, each line a thematic link between interchanges

In the augmented linkplot the central temporal diagonal of the strand recedes into an uncertain shadow beneath the richly linked, multilevel contours of thought (fig. 4). These contours seem to enable us, as the HT Writer Uther urges, "to make... our own pattern out of the disjointedness." Link Apprentice also can generate a graphic representation of hypertext links as a polar linkplot in which the x axis of this strand is mapped onto angles starting at the right and proceeding counterclockwise for one revolution, and then the y axis is mapped into distance from the center. In this kind of linkplot linear sequences appear as a central spiral, and multiple spirals reflect systematic patterns of dual creation: points of departure (Landow 1989) become radial spikes; points of arrival become circular arcs. We discern the discussion as the whirlwind, whether we see it as metaphor or measure.

Adequate and reliable discernments. We can sometimes see the whirlwind of learning take its ephemeral form in the continually replaced contours of constructive electronic text (see also "volatile" hypertext, in Bernstein et al. 1992). Yet in order to recognize, resist, appropriate, possess, replace, and deploy these contours, we need to develop what Hakim Bey proposes as the "1:1 map [of] psychotopography" (Bey 1991, 103; for the topographic, see also Bolter 1991).

0,0 6 12 18 24 30

Augmented link plot of HT Writers thread, {choices again }; each dot represents places separated by no more than 5 links

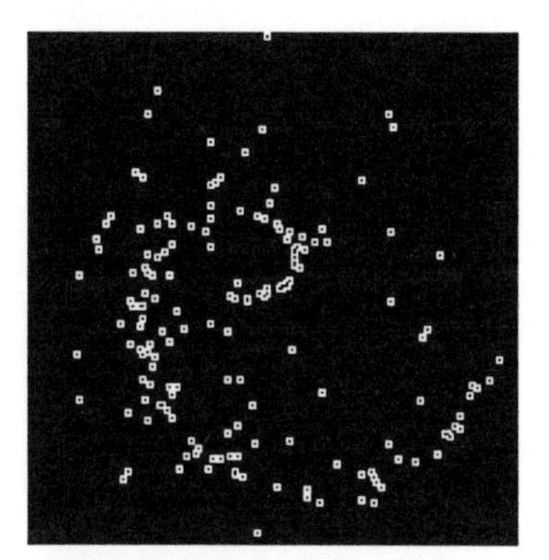

Polar link plot of the thread; linear interchanges appear as a central spiral; with spirals, spikes, and arcs of other links

Fig. 4. Two linkplots of HT Writers' thread {*choices again*}.

So mapped, Nancy Kaplan's question—"How do alternative modes of discourse... work rhetorically and ideologically?"—becomes a question of continuous discernment. "[A] 1:1 map cannot 'control' its territory because it is virtually identical with its territory," says Bey. "It can only be used to suggest, in a sense gesture towards, certain features... 'spaces' (geographic, social, cultural, imaginal) with potential to flower as autonomous zones" (103). No tool except the human psyche can approach the 1:1 mapping of autonomous reality. Even before the widespread adoption of alternative modes of discourse like hypertext (and the control it implies), we already must begin again to move, nomadically searching to discern new forms—less a virtual reality than the continually replaced contour of the real.

INTERSTITIAL

Everyone's Story Goes on without Us

Victory Garden: An Introduction

Almost from the first, right after you start thinking how there are kids in China walking upside down in sunshine while you're laying there right side up not quite asleep in the dark on the opposite side of the round world, you start to realize that the truth is that everyone's story seems to go on without us. That the truth *is* the story that goes on without us. All those people everywhere, in Buenos Aires, Bengal, Bismark, South Buffalo, Austin, and Kuwait, each in their own stories moving along side by side through time, their paths crossing from time to time, to be sure, but going on somewhere unknown to one another like the mix of extras and real people in the background of B-movie New York sidewalk scenes. If you lay there and think too long, your head starts buzzing just thinking about it, and eventually this thinking puts you to sleep, perchance to dream.

And, like most of the real truths you discover lying there alone as a kid waiting for sleep, this one never changes. Years pass, and you find yourself lying next to someone you think is your lover, look down and see her sleeping eyes rolling beneath the membrane of the lids, hear her breathing shift tempo and key, think of someone else sleeping in the dark on the opposite side of the round world, your lover or hers, think that your image was once there in her sleeping eyes, her body and yours, hers or yours or the other, once one, entwined.

It'll drive you crazy or will make you wise, thinking of all the stories and all the places they once took place and still do. For a Pynchon reader and critic it might do both. Long before he became known as the leading critic and theorist of the emerging medium of electronic writing—and before, during, and after he started writing multiple fictions in hypertext—Stuart Moulthrop was and even now remains equally well known as a Pynchon critic (the truth is that everyone's story seems to go on all at once). In fact, for fans of "logrolling in our time," this is the time to acknowledge that our stories have increasingly moved along through the same time since he became the first critic to write seriously (and not all that logrollingly, unless you think calling somebody "robotic" is a nice thing) about my hyperfiction.

Which is neither here nor there—though the truth is actually it is here, literally: my stories, yours are there, too, from time to time, in (t)his garden—neither here nor there, unless, of course, crazy or wise, you start thinking too hard of how everyone's story seems to go on without us, until buzzing dreams come scudding like allusions to the sweet silent rockets of the past. Then you think to put aside childish things and come to think of "the root sense of paranoia, a parallel or parallax gnosis, [which] happens to be a handy way to conceive of the meta-sense of pattern recognition that hypertext serves to enhance," which is how a certain proximate hypertextual Pynchonian critic and writer once described the things kids know when they lie in the dark. (The common pun intended.)

Victory Garden is the epic-length lyric of the drama of parallax gnosis (blurb that, o bestower o' the robotic!). True nonetheless, true as walking upside down in sunshine. What goes on here is the sad, sweet, synchronous truth of the day-to-day stories of all of us at war with the meaning of our lives as our lives and that meaning alike are bestowed, unbeknownst, on us. It's a love story about television, brainwaves, and course syllabi that takes place in the stormy desert of shifting electronic pixels on a screen and Chinese menu lists of punning paths.

In "Frame Up," the strangely compelling French interactive video disk game of the mid-1980s, a Candide-like character named Eddy falls asleep in a department store, where he is pursued all night both by the gang that is ripping off the store (-house of his dreams?) and by the store (house-?) detectives. At some point he retreats to the vantage of the security office, where he can view all the things that are going on at once through the video surveillance system. At this point he becomes you, your dream and his tied into a story that goes on relentlessly in the real-time (as they say) cameras you frantically switch between as—floor by floor, all at once here they come, bad guys and good guys alike—they come after you and that in the game part of the simultaneous drama you're supposed to later on edit into a coherent narrative reminiscent of the game Clue, wherein you prove your innocence and show by your news clips that it was Monsieur Moutarde in the Department *du Goods Sportif* who was guilty.

Moulthrop knows (in fact, he wrote the book about this in the same essay wherein he rolled a certain robotic log) that hyperfiction is no game. There is no edit that makes sense of it after all; rather, it is the mapping along the way that is the sense that makes the fiction (constructive, as they say, hyperfiction). He also knows that even when it seems to be a game it isn't:

> "Sorties in the hundreds," Colin Powell says at the Pentagon press briefing.
>
> So some wiseass from the press jumps up and says, "Does 'in the hundreds' mean 'less than a thousand'?"
>
> Powell (possibly amused): "I think I'll just leave it in that ballpark."...
>
> The "military analysts" are perhaps the most bizarre aspect of this whole unreal pageant, a collection of recycled generals and thinktank commanders. Color Commentators.
>
> There is the usual American analogy here, WAR=FOOTBALL. There is also the cult of expertise, the Best and Brightest Money Can Buy.
>
> CNN brings you Wolf Blitzer, too strange not to be true, while ABC weighs in with Tony Cordesman, a fellow I can only describe as Frankenstein's Correspondent (you keep looking for the cervical bolt). CBS eventually escalates all the way to General Michael Duggan, the former Air Force chief sacked some weeks back for leaking the air war game plan.

You see (and you *do* see, you do *see*, in his wondrous all-at-once world that you dip in and out of like someone mapping multifloored worlds of not cameras but, rather, interwoven narratives like witnessed dreams) that he's been there before: "In dealing with vast and nebulous information networks—to say nothing of those corporate-sponsored 'virtual realities' that may lie in our future—a certain 'creative paranoia' may be a definite asset. In fact the paragnosticism implicit in hypertext may be the best way to keep the information game clean."

Moulthrop knows. Almost from the first, right after you start thinking how there are kids in Kuwait walking upside down in sunshine while you're laying there right side up not quite asleep in the dark along the Tara River on the opposite side of the round world, you start to realize (*En-dedans Le Deluge*, as the screen goes) that the truth is that everyone's story seems to go on without us.

> Though you never comprehend the words, you come to know somehow that the PA [public address] announcement is an instruction to gather. All the families must gather up what belongs to them and join an orderly procession headed down the street or up the hill or over the bridge or across the river to the gathering place. There they must go into the auditorium and take their appointed seats, where they will wait for the dream to be over.
>
> But the dreamer, who almost invariably sees him or herself as a child, cannot join the people in the procession.
>
> Something has been forgotten—a friend, a sister or brother, a household pet or totem object. Something has been left behind.

Always something left behind. You start to realize that the truth is that everyone's story seems to go on without us. That the truth *is* the story that goes on without us. You start to realize Moulthrop knows.[1]

Quibbling-About the Author

Carolyn Guyer is one of those people who, like many women writers and most of the first generation of hyperfiction writers, seemingly have come out of nowhere.

Yet one of the contributions of feminist thinking to our time has been to remind us that coming-out-of-nowhere is a myth of power and appropriation. Those who are said to come out of nowhere in fact come from lives of real complexity, purpose, and connection. In Guyer's case her real life was one as a visual artist, a mother, and a radical educator. As a visual artist, long before her hypertexts, she envisioned an artist's book in which different colored threads were sewn through associated pages, paragraphs, images, and words, offering the reader a bright and palpable web to follow: a true hypertext.

Because until recent years she spent most of her life in Kansas, it might also be tempting to compare her to another Kansan seemingly come out of nowhere: Frank Baum's frantic and frightened projection of an idealized girlish innocence, Dorothy of the red slippers, ready to give heart to the heartless, ready to push behind the veil of male power and technology in Oz and save us all.

Carolyn Guyer isn't the Dorothy of commercial myth. Her approach to technology and power and the heart is direct, not projected. Thus, in her prospectus for Hi-Pitched voices, a collaborative hypertext writing project for women, she writes:

> Our interest in disjuncture and convoluted detail is for us an aesthetic composing rich fields of complexity. We know that being denied personal authority inclines us to prefer... decentered contexts, and we have learned, especially from our mothers, that the woven practice of women's intuitive attention and reasoned care is a fuller, more balanced process than simple rational linearity.

If Guyer has in any sense come from nowhere, it is in the sense that waves do on the lake, which situates *Quibbling.* Waves come from nowhere, or at least the horizon, in constant rhythms of difference and continuity. They keep coming and coming, washing over the edges of things, teaching us lastingness and lace as we view them closely, renewing our vision as we look for them in the distance.

"So Much Time, so Little to Do": Empowering Silence and the Electric Book

I have come to talk with you about silence.

This may seem a curious stance for a maker of hypertexts, which can seem, especially in their multimedia avatars, the noisiest of documents, all full of bugles and bangles and bright links abiding. Indeed, now that the once-upon-a-time youthful promise of the information age grows middle aged with us, more and more we seem to live much as the characters in Italo Calvino's "All at one point" (1968), a story that takes place at the "moment when all the universe's matter was concentrated in a single point" (13). A constant murmur surrounds us and becomes palpable, an atmosphere itself, as charged as the lives of those unfortunate souls we read about who dwell under high-voltage transmission wires. Surrounded by a surge of information, we spend whole days, our hair standing on end, the fillings of our teeth complaining like the red-wing blackbirds perched on the thrumming wires above us; even at the center of our cells the proteins vibrate and mutate into some new and terrible variety of information.

In the collaborative hypermedia classrooms on my campus, even at those moments when the lights are low and the machines dream of electric sheep, every teacher and learner stands at the charged center of a flow of image, word, and mutated information. Through laser disk or videotape, CD-ROM or magnetic disk, dustless colored marker or chisel point pen tip; in two-way interactive television or over satellite transmission; on CRT, whiteboard, and on paper; via internet, bitnet, phonenet, ethernet, local area net, larynx or localtalk—from here to Timbuktu—the information flows. We are surrounded and so crowded that even with the lights off and the machines asleep—their hard disks purring alpha waves—our very silence seems empowered, and yet we do not know to what end.

Like Calvino's characters, we too can say: "Naturally we were all there... where else could we have been? Nobody knew then that there could be space. Or time either: what use did we have for time, packed in there like sardines?" We have in some sense come back to the "Gate and the Center," of which the poet Charles Olson speaks, in which "the present is prologue, not the past," and we are—you are—"the only reader and mover of the instant": "You, the cause.... Get on with it" (1955, 39). But get on with what? Poised as we are in this charged silence, we can well understand the wise complaint of the wonderful Professor Wonka. For us, in this instant of unplugged silence, pressed though we may be by all this possibility, there is "so much time, so little to do."

I have come to talk with you about silence.

Obviously, this is a certain, poised silence, indeed—one that in its certainty any of us might legitimately mistake for an utterance—yet silence nonetheless. The silence at the edge of something, the silence of the "morning after dispersion," when, quoting Olson again: "If there are no walls, there are no names... and the work of the morning is methodology. How to use yourself and on what" (40). This silence I think you will recognize. It is one that, I hope, you too have admired, shared, and yet challenged as I have, seeing it in the resistant eyes of our students. I mean especially the eyes of those at risk—my Developmental Education students, e.g., or adult resumers—who have looked upon education, not with Keats's wild surmise but, rather, with the cold resolve of cultural survivors (what we might call Melanctha's eyes). It is a silence we admire as much for its resolve as for its vulnerability. Not the silence before decoding but a dissenting silence that suspects (and, in the suspicion, considers and values) the leap that links and engages us.

I have subtitled this talk "empowering silence and the electric book," not meaning to suggest a desired state but, rather, to characterize what we are already moving beyond. For it would be wrong to imagine that we might empower the dissenting silence, even with technology. Such silence is empowered already and without us—and she who keeps it knows there are more things in heaven and hell than are dreamed of in our technologies. Instead, I would suggest that we borrow from Ken Bruffee (via Polanyi) and recognize that the power that already charges and embodies silences is reciprocal. Writing the future we sing not the body electric but the body electronic, not the electric but the electronic book. What illuminates this particular bright book of life is, nonetheless, exactly what Whitman sang, the reciprocal power of our cocreative presences.

We live, you and I and our students, in what my collaborator and colleague Jay Bolter (1990) calls the late age of print. Characteristically, since Jay is a gentle and lovely man, there is a gentle and lovely pun in this term. In his phrase we are washed first with the four A.M. or fin de siècle sense of lateness, but then, belatedly, comes the undertow, the departed sense of the late, dead, lamented; what we do not want to see go and yet value all the more in the parting. In the late age of print, printed books flash through the culture, disappearing into the ether with the speed of electrons. Meanwhile, paradoxically, we are still for the moment tethered to electric books; our computers anchor us no less than the first recipients of the Jarvik hearts or the astronaut awkwardly tethered to the orbiting ship.

In our labs and classrooms most of us shepherd machines so expensive that we hardly believe ourselves when we call them tools, so expensive that we have forced ourselves (not to mention our deans and/or spouses) to forget that we, let alone our students, cannot afford them. We ration hours among our students, and allot technological upgrades to our colleagues, accounting hours and upgrades alike as actual capital within an intellectual economy.

Ironically, this scarcity expands rather than concentrates capital. It is important to realize that, as we virtualize both ownership of information and access to knowledge in our classrooms and laboratories, we move closer to a truly electronic, digital culture. When print removed knowledge from temporality, as Walter Ong reminds us, it interiorized the idea of discrete authorship and hierarchy (1982). Now, forced by circumstance to relocate the locus of information and knowledge from the object to the hour, we re-externalize ideas and make them continuous rather than discrete, reciprocal rather than empowered, contextual rather than hierarchical. Value once again becomes embedded in the social fabric rather than the distributed product. In the silence of the empty hypermedia classroom we are aware of the reciprocal power of the community of ideas that await the hour when we can plug into them. We can imagine ourselves bathed in the flow of knowledge from the space of the hour itself.

The follicle-charged cottage under high-voltage wires, the quantum tabernacle of the silent hypermedia classroom, and the roll-away book that we are tethered to like an electric heart are all scenes from the late age of print. Within these scenes and with the present as prologue, we must become what Olson calls "archaeologists of morning." We must (to echo James Clifford's *Writing Culture*) construct an actual culture from these scenes. For we have become not merely the chroniclers or custodians of but also the collaborators in a vast cultural shift. As teachers of writing in the dawning age of what Jay Bolter calls writing the mind, we are charged at our own peril to undertake a construction for which, fortunately, we are uniquely suited. The idea of the mind itself is under contention in writing the future. Great powers, both economic and political, struggle for control over writing the mind even as they lose control of its locus. "The fact in the human universe," says Olson, "is the discharge of the many (the multiple) by the one (yrself done right, whatever you are, in whatever job, is the thing—all hierarchies, like dualities, are dead ducks)" (39).

Dead ducks threaten groundskeepers, as devotees of Robin Hood or D. H. Lawrence well know. What we do with computers and writing threatens groundskeepers. A hierarchy is before anything else a search structure, a way to keep track of what you think you own, and not an arena for reciprocal power like a hypertext, a computer network, or a collaborative classroom. The peril we face is in no way minimized because I tell it as a narrative, a Robin Hood story of scenes from the late age of print. Nor is an established culture any less threatened by what manifests itself in marginal technologies such as hypertext. Margin makes meaning. Re-minding is re-forming. In her "notes from the margin" presented in 1989 Cindy Selfe reminded those attending the Computers and Writing conference that ours is a "reformist vision." The grounds that we keep (or keep at) are those of history, however marginal or powerless our meaning-making professions—teacher, writer, hypertext maker—may sometimes seem.

Writing the mind, like any writing, is a process; a change of mind changes how we think of ourselves in time. The nature of mind must not be fixed either in time or in silicon. It is a conversation we must keep open. Hypertext links are no less than the trace of such questions, a conversation with structure. Like the conversation of our teaching, it is authentically concerned with consciousness rather than information, with creating knowledge rather than the mere ordering or preservation of the known.

I have come to talk with you about silence.

Here I want to make a confession. The words of the recurrent verse I'm preaching have for me an actual substance and a palpable extension, which is I think borne of my work with hypertext and writing. I take seriously what might seem a metaphor in my hypertext fiction, "afternoon," the idea of words that yield on account of their texture. It is as if I can reach somewhere into my own mind and history and press against this sentence (perhaps "express it" is the right image, as words from the child haltingly come, or as milk from the breast). What I have in mind here is topographic writing, or "spatially realized topics," referring to "signs and structures on a computer screen that have no easy equivalent in speech." Or at least not in Western, pre-hypertextual speech.

Not long after I first met Stuart Moulthrop, I remember asking him, in all seriousness, "Do you ever find yourself wanting to press the words on a page of a book you're reading to see what's behind them? or wishing you could probe the words as you're listening to someone?"

In all seriousness, Stuart answered, "No."

Even so, he knows what I mean. Or so I console myself when he and Nancy Kaplan warn against our too easily dismissing recently published absent-minded musings and soi-disant research about graphical writing (Kaplan and Moulthrop 1990). For at very least (and likely at best) they involve a recognition of real change, a change that involves exactly this sense of the language as yielding object. Not that the change is coming but, rather, that it has come, and we are its expression.

So, when I say the first phrase in the verse "I have come to . . . ," it first, naturally and reciprocally, yields for me as if something in space, softening, opening. Behind it there is a whiff of lizards and parlor lace, something as alien and high-minded as a two-year-old rerun of William Buckley's "Firing Line," as in "I have come to *love* quoits . . ." Suddenly, I recall Robert Creeley's meditation in which he turns what once for me seemed a similarly high-minded phrase to something accurate: "I'm given to write poems," reads his first line and the essay's title, "I cannot anticipate their occasion" (1970).

My kind of Irish is neither lace curtain nor shanty but, instead, what we can call pig sty (getting half the phrase and the second initial consonant right). We never quite had the leisure or the urge to be given to anything, but when I read Creeley's essay I knew what the social construction of this phrase could mean. More, I knew I could own it. So too, the also high-minded—not to say High Church—aspects of the formulation "I have come to . . ." are suddenly something I can recover and make my own. In fact, I begin to wonder if my own rejection of such formulations wasn't simply a reflection of my working-class resentment of those who had time to think such things in such fine ways.

To reclaim phrases from classes characterizes the reciprocal power of the electronic book. Freed from hierarchy to multiplicity, with the present as prologue, having so much time and so little to do, we and our students might become the bastard offspring of Borges's Pierre Menard and the living books of Truffaut's version of *Fahrenheit 451*. We might possess processes that we were only once the property of. The groundskeepers might enjoy the landlord's (or lady's) favors, so to speak, and come, like the walrus, to think of many things in such fine ways.

Not that the change is coming, but that it has come, and we are its expression.

When I say "I have come to . . ." I also have in mind here the intellectual journey. Much of it has been the kind of accretive, accumulative, layering knowing that Fred Kemp and others describe as the polylogue nature of electronic texts. My own transition from what I once believed to what I have come to know—from print to topography, from book to "e-text"—seems this kind of sedimentary layering of voices and occasions of the sort that hypertext theorist Randy Trigg calls "collaborative annotation" (Trigg, Suchman, and Halasz 1986). In polylogue the geology of mind is successively overwritten and ultimately cannot be separated from the aggregate; outside polylogue, change etches the shocked singularity of fossils into stony shadows of themselves. When Jay and I began our six-year machine-based conversation about what we did not then know to call hypertext, I wanted to create a book that would change every time you read it. Yet I envisioned a disk tucked into a book, and not the opposite. It took a group of people to show me my folly: Jay, Jane Yellowlees Douglas,

Stuart, Nancy Kaplan, John McDaid—and now I cannot separate myself from the aggregate without crumbling to dust. Even so, I remember struggling against the electronic book.

"But you can't read it in bed," I said, everyone's last-ditch argument. Now in the year when Sony showed Dataman, a portable, mini-CD the size of a Walkman, capable of holding 100,000 pages of text, I am amused to see a discussion on the Gutenberg computer network that wants to move the last ditch a little further. The smell of ink, one writer suggests; the crinkle of pages, suggests another. Meanwhile, in far-off laboratories of the Military-Infotainment Complex (to advance upon Stuart Moulthrop's phrase), at Disney or—yea, verily, even unto—Apple, some scientists work on synchronous smell-o-vision with real-time simulated fragrance degradation shifting from fresh ink to old mold; while others build raised-text touch screens with laterally facing windows that look and turn like pages, crinkling and sighing as they turn. "But the dog can't eat it," someone protests, and, smiling silently, the scientists go back to their laboratories, bags of silicon kibbles over their shoulders.

We are in the late age of print; the time of the book has passed. The book is an obscure pleasure like the opera or cigarettes. The book is dead—long live the book. A revolution enacts what a population already expresses: one hundred thousand videotapes are sent to a television show about home videos. What will it take to convince us? Perhaps the however-many thousand who regularly write e-texts on networks?

Coming generations will likely see the body of research about computers and composition as history written in a curiously nervous style. Consider, e.g., research—some of it quite good—that measures how readers and writers plan and revise electronically against how they did in print. What is it but longing in an age when more and more writers at work do so electronically? when print is increasingly a fetish reserved for drafts, while the real writing is transmitted electronically? when few ever plan, revise, or author the whole of a document individually?

And what will coming generations make of the kabbala of workbench readabilities (what the Ed School types now want to call considerateness) and the theosophical metrics of machine-based style checking? Perhaps they will account for our ciphering as more of the superstitious quantifying of an age of opinion polls: how we once thought we could measure—on a single surface, mind you—the multiple enfoldings and disclosures possible within an electronic communication, in which a sentence can wind through distant pages, assume multiple voices, and shift as the reader wills or as she interacts with others.

Yet those who would write such nervous histories and superstitious quantifying should hurry. For awhile, to be sure, both the rapaciousness of the technology industries and the backwardness of our institutions will buy them time, but the change will be swift, for it is already upon us. The cultural shift to e-text may not be so radical as the recent Gutenberg net suggestion that all libraries might suddenly go on-line, but it will probably be as radical as the shift from print to tube and surely will be as swift as the transition in the audio world from LP to CD.

Not that the change is coming, but that it has come, and we are its expression.

It is happenstance that on either end of my town, like compass points, there lies a mausoleum of books. On the north end of town lies a great remainder warehouse promising 80 percent off publishers' prices. Inside it row upon row of long tables resemble nothing less than those awful makeshift morgues that spring up around disasters. The tables are piled with the union dead—the mistakes and enthusiasms of editors, the miscalculations of marketing types, the brightly jacketed, orphaned victims of faddish, fickle, or fifteen-minute shifts of opinion and/or history. Here an appliance is betrayed by another (food processor by microwave); a diet guru is overthrown by a leftist in leotards (Pritikin by Fonda); and every would-be Dickens seems poised to tumble, if not from literary history, at least from all human memory (soon gangs of Owen Meanies will leer and lean against these faded Handmaidens of Atwood).

When I first saw such a warehouse—forty miles east of our spare parts, Bible Belt midwestern town, in what we outlanders think of as wonderful Ann Arbor—I thought only a university town could sustain this. When the same outfit opened up in our town, and the tables were piled with towers of the same texts, I knew this was a modern-day circus. Ladies and gentlemen and children of all ages! here come the books!

Meanwhile, at the opposite pole in the second mausoleum a group called the Friends of the Library sell off tables of what shelves can no longer hold. One hundred years of Márquez is too impermanent for the permanent collection of our county library, but so too—at least for the branches that feed pulp back to this trunk—so too is the Human Comedy, so too are the actual Dickens or Emily Dickinson. Here the book must literally earn its keep.

Both the remainder morgue and the friends of the library mortuary are examples of production and distribution gone radically wrong. At first these seem depressing scenes, yet they offer us (as archaeologists of morning) more than enough to build a culture upon. Books—and films and television programs and software—have become what cigarettes are in prison, a currency, a token of value, a high-voltage utility humming with options and futures.

The intellectual capital economy has to some extent abandoned the idea of real, material value for one of utility. This abandonment is not unlike the kind that leaves so-called worthless condos as empty towers. Ideas of all sorts have their fifteen-minute Warholian half-life and then dissipate, and yet their structures remain. We might, like urban squatters, delude ourselves into thinking that, because the cultural economy has abandoned them, we can move in. If we do so, we will be sitting in the sand under the high-voltage wires. They have long ago stopped making real buildings in favor of virtual realities and holograms. The book, no less than the sitcom, is merely a fleeting, momentarily marketable, physical instantiation of the network.

It is here, within the network, that we should continue to build, here that we should put our inhabiting energies. In the days before the channel zapper and modem port we used to think *network* meant the three wise men with the same middle initial, NBC, ABC, and their cousin CBS. Now we know that the network is nothing less than all our voices (in classroom, hypertext, or electronic life) put before us for use. Here in the network what makes value is knowing how to use yourself and on what. As in the reciprocal silence of the empty hypermedia classroom, our networks build locally immediate value that we can plug into or not as we like. Here the Wonkian proverb yields its meaning: "so much time," because the network redeems time for us; "so little to do," because what is of value here is what can be used. Once transmitted our voices have the power to shift values everywhere, instantly, individually, as we may think.

Not that the change is coming, but that it has come, and we are its expression.

I claim to have come to talk with you, not to you. No matter how singular, my voice is many in the polylogue. This marginalizes me in a reciprocal relationship with you and each of you in turn with the rest of us. The network, the electronic book, the truly digital culture, links voices at the margins. What puts us in peril is how powerful we are at this margin. We are powerful because we know and teach that the value of something—especially the value of text—resides, to use some old, comfortable terms, in process rather than product. Or, to put it differently, our texts are in action.

I take the phrase from Bruno Latour's revolutionary *Science in Action* (1989) in which he notes that technical literature is "hard to read and to analyse not because it escapes from all normal social links but because it is *more* social than so-called normal social ties" (62). I.e., the text is a network available for use, and to truly use it we must either create it for ourselves or contribute to its demise. For Latour there are "only three possible readings [and] all lead to the demise of the text":

> If you give up, the text does not count and might as well not have been written at all. If you go along, you believe it so much that it is quickly abstracted, abridged, stylised and sinks into tacit practice. Lastly, if you work through the author's trials, you quit the text and enter the laboratory. (61)

The electronic book is left to enact either an orthodoxy or the laboratory. In this sense it *is* knowledge in action (or what Eco calls "work in movement"). Elsewhere I have distinguished two kinds of hypertext according to their actions (Joyce 1988). Exploratory hypertext, which most often occurs in read-only form, allows readers to control the transformation of a defined body of material. Its demise is into the stylized abstraction of orthodoxy, the second of Latour's alternatives, and as such is inherently dangerous. To the extent that exploratory hypertexts give us the illusion of control, we are apt to believe too much in them and so risk sealing others' power over us through our tacit practice. Control itself can be made to seem a product.

The nature and control of products is absolutely at issue in hypertext. Stuart Moulthrop suggests that to the extent that hypertexts let the power structure "subject itself to trivial critiques in order to pre-empt any real questioning of authority... hypertext could end up betraying the anti-hierarchical ideals implicit in its foundation." The critique becomes orthodox and seems satisfying in itself: witness Gary Wills's recent front-page so-called critique of Dan Quayle in *Time* magazine (1990).

The present hypertext is largely exploratory and so far manages to seem successful in what ultimately is an impossible attempt: the preservation of hierarchy through managed individuality and the illusion of control. The present hypertext is thus consumerist at base. What we consume is obviously not the information but, rather, the occasion it presents us with. To remove hypertext programming—in the double sense of instructed machine behavior and information content—from consumer culture, we must move from product to construct.

Constructive hypertext quits the text and enters the laboratory. Constructive hypertext requires a capability to create, change, and recover particular encounters within a developing body of knowledge or writing process. Like the conference or classroom or any other form of the electronic book, constructive hypertexts are versions of what they are becoming, a structure for what does not yet exist. For silence.

Finally, we have come to the last yielding measures of my recurrent verse. I have come to talk with you about silence. Once we are in the laboratory we are about silence. I like the word *about* because it suggests both marginality ("I'll just walk about...") and focus ("I know what this is about")—because it suggests both proximity and concern. We are always about silence in what we do. We should not congratulate ourselves, however, for either our proximity or our concern. Whether the laboratory is the hypertext, the networked classroom, or the computer conference, one of its possible functions, as Latour reminds us, is to muffle (if not crush) dissent:

> "You doubt what I wrote? Let me show you." The very rare and obstinate dissenter who has not been convinced by the scientific text, and who has not found other ways to get rid of the author, is led from the text into the place where the text is said to come

> from. I will call this place the laboratory... the place where scientists work. (1989, 64)

In the eyes of the dissenter we have come full circle to the eyes of those at risk, the cold resolve of cultural survivors, to Melanctha's eyes or our eyes. O what do they fear? we ask ourselves. O what do we fear?

Not that the change is coming, but that it has come, and we are its expression.

What frightens us, I think, is the loss of the analog, the ascent of the digital. A student of mine writes in his journal (it has been composed on a computer and printed in three-column format in the guise of a newspaper): "With a record I could get some sound off it even if I had to use a bamboo needle and a klaxon speaker like we did once in Nam. With a CD there's nothing I can do to get at those zeros and ones." I did not think then to suggest to my student the irony in this parable of war against the unseen.

The digital is the topographic, the zeros and ones that mean by merely being, what Latour calls the "new object." What we fear is that reality itself will be altered in a way we cannot get at—no matter that the wave forms of analog sound were once as much a new object, once as green a thing as the new bamboo shoot. "Laboratories," says Latour, "are now powerful enough to define reality.... [R]eality as the latin word *res* indicates, is what *resists*. What does it resist? *Trials of strength.* If, in a given situation, no dissenter is able to modify the shape of the new object, then that's it, it *is* reality..." (1989, 93).

The new writing—the topographic, truly digital writing—even now resists attempts to wrestle it back into analog or modify its shape into the shape of print. Its resistance is its malleability. We cannot get at it because it will not stand still. As Richard Lanham puts it, "Digitization both desubstantiates a work of art and subjects it to perpetual immanent metamorphosis from one sense-dimension to another" (1989, 268). Ted Nelson's definition of *hypertext,* i.e., "non-sequential writing with reader controlled links," stops short of describing the resistance of this new object (1987, v). For it is not merely that the reader can choose the order of what she reads but also that her choices, in fact, become what it is.

Here is a nest of paradoxes. If we wish to know the text our reader has written through her choices, we will have to reciprocate. She can only know by our choices what distinguishes her own. We construct the new writing by our choices, but we only come to know what we have written by understanding the choices of others. Individual power is reciprocal. In the shift from the analog to the digital, culture discloses itself as choices, the topographic discloses itself as polylogue. To get at the zeros and ones that are the music of our spheres is to enter the network. The new writing is, like Latour's scientific writing, *more* social than so-called normal social ties. We understand from the third person what we have written in the first person but only in the process of reading the second person.

Current disputes about graphical–versus–command line interfaces evidence these paradoxes as trials of strength. Graphical interfaces seem at first to rob us of personality and lock us into shadows of a third person on the cave wall before us. We are the arrow among the icons, just one among all the meaningful things projected there. Command line interfaces put us into the first person; we seem to assume a personality thought to control the machine. We are the word, the light, and the analog, or so we think.

They would like us to think so. A constant murmur surrounds us and becomes palpable, an atmosphere itself, as charged as the lives of those unfortunate souls who dwell under high-voltage transmission wires. It is the hurly-burly of the vast amusement park, the ultimate infotainment. Over this Gate and Center, the Disneyites raise a gaily colored banner upon which the strong argument of Cognition is writ large: it says that the only way to understand and navigate all this, the only way to sail among these continents of knowledge, is to purchase the alternate you they have kindly constructed for you—not what William Gibson calls "simstim," but, rather, what he calls an AI. How you think is known, quantifiable, recognizable, even unto its serendipity and opportunism. No longer any need to be just one among all the meaningful things.

Over against the Disneyite is the Wonkian network of elevators that go through the roof, electronic gobstoppers and other new objects. Here there is no hurly-burly, only the reciprocal silence of so much time and so little to do. Because there is no single gate, there is no gaily colored banner. Instead, a hundred thousand tiny notes, scraps the size of fortune cookie fortunes, flutter in the wind like Rushdie's "miraculous squadrons... of... butterflies [which] had become so familiar as to seem mundane" (1989, 217). Each in its own tongue advocates the immediacy of another view, the possibility of giving oneself over to another as another, rather than to the other in the guise of oneself.

The real trials of strength before us finally involve our ability to champion the Wonkian over the Disneyite, the constructive over the exploratory, process over product, reciprocal power over empowerment, network over programming. We are called in trials of strength to champion the individual reality of others' ideas against those who would claim to find another reality for them. The struggle we face, the peril we are in, literally involves the nature of reality and the conservation of empathy, analysis, and disinterestedness. We have learned from the resistant eyes of our students how to conduct ourselves in this struggle. In such a struggle it is an advantage to have lived so long at the margin. Our advantage is our marginality, the resistance that makes of us the new topographic object, the new writing.

I have come to talk with you about silence. Not that the change is coming, but that it has come, and we are its expression.

A Memphite Topography: Governance and the City of Text

Imagine a city of text. Because this essay evolved from a talk I first gave there, I think of Memphis, Tennessee and its etymon, the founding city of the third dynasty, the city in which, according to the Memphite Theology, the order of all being emerges from "what the heart thinks and the tongue commands."

"Last time I saw you," the poet Lucille Clifton writes,

> ... was on the corner of
> pyramid and sphinx.
> ten thousand years have interrupted our conversation
> but I have kept most of my words
> till you came back.
>
> (Clifton 1987, 71)

Imagine. O yes, it could be Memphis, or even Lucille Clifton's Depew, this self-proclaimed Dahomey woman's birth home, a small Polish steel-making city outside Buffalo, my birth home. Imagine a city of text.

Or it could be Itil, the capital city of the Khazars in Milorad Pavic's novel *Dictionary of the Khazars,* in which "there is a place... where when two people (who may be quite unknown to each other) cross paths, they assume each other's name and fate, and each lives out the rest of his or her life in the role of the other, as though they had swapped caps" (1988, 73).

Imagine a city of text. Pharae, perhaps, the city of the oracle of Hermes in which those who consult the god "are granted their oracular responses in the first words that they hear on leaving the market place." Imagine wisdom overhead beyond the portals of the shopping mall (Graves 1955, 179).

Or Gloucester, the city that is formed by the hero of Charles Olson's epic Maximus Poems as he addresses it:

> I speak to any of you, not to you all, to no group...
> ...Polis now
> is a few, is a coherence, not even yet new (the island of this city
> is a mainland now of who? who can say who are
> citizens?
>
> Only a man or a girl who hear a word...
>
> (Olson 1960, 11)

Imagine an envoy from a city of text: from Memphis, Depew, Buffalo, from Itil, Gloucester, and Pharae to this new Memphis. Like you, I am interested in the city as a place in which language transforms us. Like you, I am interested not merely in future tense, if you will, but future text. I am interested in the text as a place of encounter in which we create the future. Yet like any envoy, I bring only troubling questions and a history of intentions on my back.

In Pavic's novel "one of the envoys had the Khazars' history and topography tattooed on his body" (73) and "lived... like a living encyclopedia of the Khazars, on money earned by standing quietly through long nights":

> He would keep his vigil, his gaze fixed on the Bosporus' silver treetops, which resembled puffs of smoke. While he stood, Greek and other scribes would copy the Khazar history from his back and thighs into their books. (77)

Not very good at standing still and never one to keep quiet, I have nonetheless for some time now been able to keep my gaze fixed on what seem to me at least to be the silver tops of trees but which others may take for mere puffs of smoke. As a writer of hypertext fictions and cocreator of a hypertext system for readers and writers, I began some years ago with a dream of a novel that would change each time one read it, that would not stay the same from one reading to another each time you put it down. In the course of realizing that dream, I have had the corresponding good fortune of being made to think from behind, to see myself through by a group of friends who have become for me a city, in Olson's sense of polis as a coherence. These friends have read my intentions from the back, asking difficult questions about authorship in hypertext, about the authority of the reader, about multiplicity and control, about the future and past of the city of text. Questions that invite you to embody them yourself, to give them heart and tongue.

The original Memphite Theology was also in some sense reading from the back. Memphis was a city of text that gave itself the task of reconciling the ascendency of Ptah against the Atumic creation myth. That myth held that Atum brought the gods into being by naming the parts of his own body. The Memphite Theology breathed possibility back into a world exhausted by its own making, giving the world back the heart and tongue that come before the naming.

We, too, live in a world increasingly exhausted by its own making. Elsewhere I have described our exhaustion as a "constant murmur [that] surrounds us and becomes palpable, an atmosphere itself, as charged as the lives of those unfortunate souls we read about who dwell under high-voltage transmission wires." The city of text itself pulsates within a charged vortex of invisible power. How can we read it, we wonder, when it so shifts on us, when we resonate with it, and when, writing us, it writes with us? Previously, I tried to argue that—even now in its peak demand as a preferred political/pedagogical term—we ought to reject the notion of empowerment, especially, as Nancy Kaplan puts it, "empowerment as a transitive verb, i.e., 'I empower you.'" Or, failing that, we ought at least to extend the term to recognize the reciprocal power of silence in the face of the charged and constant murmur of language in the future tense, the future text.

I am not now so certain that the connection between empowerment and reciprocity is as simple as valuing the one or the other term. "What seems more compelling," quoting the poet and theorist Charles Bernstein, "is to understand (be troubled by) the situational dynamics of characterization" (1986, 385). In future tense we see ourselves in troubled silence out of which we create, in Bernstein's terms, "a music of contrasting characterizations" (446). It is a music made of alternating currents, a music that, building on Pierre Boulez's distinctions between smooth and striated space-time, Carolyn Guyer suggests is formed of the rhythmic interplay of acceptance and control:

> The process of creating ourselves always involves two polar events: Acceptance and Control, that is, *occupying without counting,* and *counting in order to occupy.* One is not preferable to the other;

> rather, neither exists without the other, which means that the only thing we can truly be interested in is the complex mixtures of the two, how they proportion themselves as they move through each other. (1992, 3)

The interplay of acceptance and control, the music of contrasting characterizations, echoes in the complex mixtures of Gertrude Stein's Melanctha, whose fictional wanderings mapped the city of text. Her acceptance of the city rhythms of "always knowing what it is I am wanting" ultimately clashes with the controlling urges of Jeff Campbell, her lover and would-be mentor, a troubled doctor of medicine and mind, who, Stein tells us, "took a book to forget his thinking, and then as always, he loved it when he was reading" (1909, 170). "No, Jeff Campbell," says Melanctha:

> It certainly ain't that way with me at all the way you say it. It's because I am always knowing what it is I am wanting, when I get it. I certainly don't never have to wait till I have it, and then throw away what I got in me, and then come back and say, that's a mistake I just been making.... It's that way of knowing right what I am wanting, makes me feel nobody can come right with me, when I am feeling things, Jeff Campbell. I certainly do say, Jeff Campbell, I certainly don't think much of the way you always do it, always never knowing what it is you are ever really wanting and everybody has got to suffer. (183)

Melanctha's "way of knowing right what I am wanting" sings Stein's deep prophecy of woman and the city in the dawn of an age of instantaneous life. The controlling book versus the acceptant text: some of us do not suffer perturbation well. Jeff Campbell is so frightened by Melanctha's charged world wandering that, thrown into simultaneous and contradictory actions, "he rose up, and thundered out a black oath, and he was fierce to leave her now forever, and then with the same movement, he took her in his arms and held her." But this embrace cannot hold her, cannot read her: "No, Melanctha, little girl," he dismisses and diminishes her, "really truly, you aint right the way you think it..." (184).

Imagine a city of text. Who will rule it? What will we call it? In what way shall we write it? How shall we read it? A troubling silence falls over us and yet we sense we are in movement. Where? We need to think from behind to see ourselves in movement. "In order for the third body to be written," says Hélène Cixous, "the exterior must enter and the interior must open out" (1991, 54). We need to see ourselves through, at depth and engaged, within the historical scene not confronting it, authoring the text of one's own future, knowing right what we are wanting, projecting and not projected upon. "What is open is time," says Cixous, "not to absorb the thing, the other, but to let the thing present itself" (63).

In a poem Olson characterizes this kind of knowing in terms of the dancer who can envision her own back as she moves in space. Elsewhere he imagines it as a slug of type, the object that knows its own form mirrored and reversed. In his essay "Proprioception," which names this kind of literally knowing through oneself, he says, "The advantage is to 'place' the thing [i.e., the unconscious], instead of it wallowing around sort of outside, in the universe." "The gain," he suggests later, "is to have a third term, so that movement or action is 'home'":

> Neither the Unconscious nor Projection (here used to remove the false opposition of "Conscious"; "consciousness" is self) have a home unless the DEPTH implicit in physical being, built-in time-space specifics, and moving... is asserted, or found out as such. (Olson 1974, 17, 19)

A third term, the third body: the task of our troubled silence and our wild surmise is to find a home in the constant murmur and movement of the city of text. It lies before us as a network of light, an ocean deceptively pacific, roiling, and itself ever moving through time. Accounting the "depth implicit in physical being" was in some sense the impetus of the Memphite Theology, which set out to reinhabit a world flattened by its own articulation. "What the heart thinks and the tongue commands" provided "built-in time-space specifics" for a world in movement, a world re-embodied. "Medium my body, rhythmic my writing," says Cixous.

Proprioception is the body's knowledge of its own depth and location, its internalized perspective of "how to use oneself and on what," in Olson's phrase. Yet graphical computer interfaces, hypertexts, virtual realities, and other instances of what Haraway calls the "couplings between organism and machine conceived as coded devices" serve to externalize the internal. Such varieties of technological experience insist upon the permeability of the "self" and the "what" and the uses made of each. Cyborg consciousness invites us to turn proprioception outward beyond Haraway's "crucial boundary breakdown" of organism-machine to a place where we take "pleasure in the confusion of boundaries and responsibility in their construction."

At first electronic writing appears to threaten time-space specifics, the essential "thisness" of the body of text. Yet whole cities are painted in it, and even the smallest burg (O, my Jackson!) pulses as radiant as the Ginza when viewed as a text. Ultimately, electronic writing confirms our proprioceptive sense of the textness of the body. "I don't 'begin' by 'writing,'" writes Cixous: "I don't write. Life becomes text starting out from my body. I am already text. History, love, violence, time, work, desire inscribe it in my body..." (52).

Even from the most common virtuality, i.e., the framed antimatter on every wall—the everpleasant TV, the constant tube—there pours forth a shimmer of transcendent text. A newscaster reporting the decline of literacy never considers the transitory nature of the text that captions that decline, its projected letters shimmering lime and melon upon the simulacra of the chroma key floating in the air just beyond her shoulder. The lime-melon banner graphic is most likely a piece of subscription video clip-art graphic up- and down-linked from the network in New York, where hours before a young network graphic artist has used an electronic "airbrush" to limn a caption writer's word-processed half-sentence e-mailed earlier over the network to her electronic tablet. What the newscaster mouths, most likely without considering it except as light before her, is most likely equally word-processed, e-mailed, and transitory, a script that steps its way up the ladder of virtuality from terminal to teleprompter to unheeding ether without benefit of intervening paper.

A citizen taping the broadcast has used a Visa card to purchase the videotape. (Or she has used an ATM card to withdraw government hard copy: measured futures given currency in somewhat gloomily illustrated green-inked visages of great, dead white men whose texts are never read.) In either transaction the citizen executes a contract she likely has also never read nor will, while the plastic card executes computer code that the sale clerk need not read and that the teller machine cannot.

Later, the videotaped newscast may be summoned within the text of a hypermedia system, poured out from a video spigot (words that, capitalized and trademarked, name a product, a printed circuit, a city of silicon and gold), the image flowing along the also silicon avenues of the dark city within the case of the computer into some dark sink, a reservoir of possibility. There, where no human ever has walked, a third-generation cousin of the original graphic—frame captured and optically scanned—may rise up from the dark water like Mothra, the original picture of its text now turned back into pixels of "actual" coded ASCII text and the recorded audio digitally decoded (even scored and transposed onto musical staffs, ready to be played, sampled, syncopated).

Broadcast seed, though cast into the wind and falling on the sands, nonetheless relentlessly thrusts up, sprouting back into text in much the same way as sown dragons' teeth once brought forth races of gods. The existence of any atom of literacy—text itself, the word *thisness*, etc.—depends upon our interaction with it. As in the primordial word of the ancients, or the paranoid's dream, each letter threatens or promises to give way to some mutant foliage, some elemental wave leading outward from it.

In laboratory hypermedia systems, even the letter itself, like *e coli*, subdivides into the individually significant pixel; even the void between letters, or the space between the perfect incisors of the newscaster's smile, can be filled with the invisible traces of a link to something else. Hypertext, I have noted previously, is the revenge of the text on television. The late age of print is not the age of multimedia but, rather, of the multitext, not the age of hypertext but of the multitext. William Carlos Williams's poetic dictum, "No ideas but in things" is instantiated in the music of contrasting characterizations embodied in things as they are.

What results is something of a quantum theory of language, in which the text as quantum serves as the "particle mediating a specific type of fundamental interaction," to borrow the American Heritage definition (a borrowing that, of course, may be done literally in an electronic text). By *placing* texts in their topography we re-externalize ideas and make them continuous rather than discrete, reciprocal rather than empowered, contextual rather than hierarchical. Value once again becomes embedded in the social fabric rather than the distributed product. Imagine a city of text as a litany of placements:

> the place where the heart thinks and the tongue
> commands;
> the place where ten thousand years have
> interrupted our conversation but we have
> kept most of our words;
> the place where people cross paths and assume
> each other's name and fate;
> the place where wisdom is overheard outside
> the shopping mall;
> the place where a few is a coherence;
> the place where a constant murmur surrounds
> us and becomes palpable;
> the place where the whole sounding gets heard;
> the place where we are always knowing what it
> is we are wanting;
> the place of polar events where the only thing
> we can truly be interested in is . . . how they
> proportion themselves as they move
> through each other;
> the place where perception of an object means
> loosing and losing it.

The litany of our encounters performs a music of contrasting characterizations; placement places us within a narrative. "The Khazars," Pavic writes, "imagine the future in terms of space never time" (1988, 145). In this way we embody our own intentions. I place myself as an envoy to you and try very hard to keep still for you, my gaze fixed on what I think to be silver tree tops.

Yet it may be the towers of a new city that I see: Memphis perhaps, a city that, because it is networked, is removed from the hegemony of The Networks; which, because it is kaleidoscopic and multiple, can neither be focused into a single white light nor broken and separated into individual strands of the color spectrum. Instead, a dapple of text falls upon the new city as light, laps against it, washes it. The city of text grows from our broadcast seed; rhizomic in Deleuze and Guattari's sense, "more grass than a tree," space never time. Imagine this city of text. Who will rule it? What will we call it? In what way shall we write it? How shall we read it?

Topographical writing enables "signs and structures that have no easy equivalent in speech" (Bolter 1991, 25). Power relationships, canonical distinctions, notions of polarity, and so on show themselves as light on a troubled surface, the true bent of language. As we near the new city, we are, to use the stunning characterization and evocation of Emily Dickinson from Susan Howe's book, *My Emily Dickinson,* "Nearer to know less before afterward schism in sum." In that sentence is a story of knowing and unknowing. "Narrative," Howe characterizes both her own book and Dickinson, "expanding, contracting, dissolving. Nearer to know less before afterward schism in sum. No hierarchy, no notion of polarity. Perception of an object means loosing and losing it" (1985, 23).

No hierarchy. Grass that grows like wild fire. The city of text is formed on a belief that we are the outcome of stories we tell ourselves as we come to know new things. So, a story:

On a bright, hot Autumn afternoon in the Republic of Kansas some years ago, on precisely the eve of the cold snap after which all the leaves fell, we walked for this one last afternoon among the saffron and vermillion leaves of young oaks, the philosopher and I talking of hypertext. "You can't let the students change Plato, can you?" he asks, "Surely you can't let them do that."

There: the issue is engaged, the troubling question topographically embodied. What do you say, Phaedrus?

I could have said no; I could have said yes; I could have asked pleasantly, "Which Plato?"

The one who asked the question was a true philosopher, a sincere and genuine teacher, a tough mind (walking with him was wrestling angels), and wise enough to know that his own question described a possible future.

You might say, "Well, of course, it isn't possible for the students to change Plato." Yet the philosopher thought it was, and as an envoy to the city of text I do also.

Ted Nelson, who coined the term *hypertext* in the 1960s, later defined it as "non-sequential writing with reader controlled links" (1987, v). This characterization, I think, stops short of describing the resistance of this new object. For it is not merely that the reader can choose the order of what she reads but that her choices in fact become what it is. Let us say, instead, that hypertext is reading and writing electronically in an order you choose; whether among choices represented for you by the writer or by your discovery of the topographic (sensual) organization of the text. Your choices, not the author's representations or the initial topography, constitute the current state of the text.

What, in our prairie land peripeteia, the philosopher saw was that the trouble with hypertext at any level is that it lets you see ghosts; it is always haunted by the possibility of other voices, other topographies, others' governance. Of the responses open to me, my best choice might have been "Which Plato?" The original collation of Thrasyllus? Else whose edition or attribution? Proclus? Ueberweg? Praechter? Jaeger? Taylor? The Plato of my mind? Of James Joyce's? Of Marx? Of Susan Howe or Susan Sontag? Of Lucille Clifton? Of Erin Mouré?

Until you are a graduate student (and often even then and always after) in a culture, we do this work for you. As a culture, we would like to think this a neutral decision, unladen, decontextualized, removed from issues of empowerment, outside any reciprocal relationship. In fact, hypermedia educators frequently advertise their stacks by featuring the fact that the primary materials are not altered by the webs of comments and connections made by students. This makes it easier to administer networks, they say.

Or to prevent the appearance of ghosts. At present, institutions of media, publishing, scholarship, and instruction depend upon the inertia of the aging technology of print, not just to withstand attack on established ideas but also to withstand the necessity to reestablish their own ideas. We prevent the student from changing Plato in our systems so as to avoid having to engage ourselves in actively maintaining the intellectual apparatus of Plato. We do not want to have to rethink Plato. And the less we do not, the more and more it seems that a sufficiently strong attack could unseat Plato.

Luce Irigaray notes that discussion of the canon "seems to presuppose that all that is decided as of now, that there will only be the past in the future, that the female and male readers living beyond the twentieth century will have no part in deciding what will define the twentieth century canon... that there will be only one canon and it will have only one content" (1993, 57).

Like Cachulain, they lose who battle waves on the shores of light. The book is slow, the network is quick; the book is many of one, the network is many ones multiplied; the book is dialogic, the network polylogic. "If there's only one [canon]," says Irigaray, "it would represent an immutable framework of language": "You seem to ignore the fact that there are several languages and that they evolve." In the network if you can't change Plato, you clone him and make your changes multiple: My Emily Dickinson, My Plato. Who will maintain Plato if not this woman, our student on the network? How will she maintain him except to transform him?

When I first began thinking of these things I didn't think real transformation was possible in hypertext. In an essay called "Selfish Interaction," I wrote:

> The attendant "pleasures"... involve confirmation of interaction... what is confirmed are limits, or the traversal of boundaries... between the act of reading in a certain way and the text which anticipates and signifies successful reading in that certain way. Thus we are unlikely to write intertextual, parallel, or detached versions... because we selfishly seek confirmation of our alternative choices within the text.

Lately I have tried to think how the topographic (sensual) organization of the text might present reciprocal choices that constitute and transform the current state of the text. How can the reader know what ghosts her choices evoke, the places where they walk, or how they proportion themselves as they move through each other? How can she know what she is "always knowing what it is [she is] wanting, when [she gets] it?"

We must make certain only that the city of text in Irigaray's phrase "contribute[s] to the transformations in the forms and contents of discourse." We must make certain only that it is consciously unfinished, fragmentary, open, that it becomes a locale for Haraway's cyborg, she who "skips the original unity" and is "resolutely committed to partiality, irony, intimacy, and perversity."

John McDaid's hyperfiction *Uncle Buddy's Phantom Funhouse* is an electronic world of notebooks, scrap papers, dealt but unplayed Tarot cards, souvenirs, segments, drafts, and tapes, unfinished in the way that death unfinishes us all. In a journal within an early hyperfiction Carolyn Guyer asks herself, "What are you gonna do about the reader?" then lists ways to "try to create a flexing (fluxing) text" for herself and her reader "who, after all, is writing the thing herself in the reading." Yet, "how bloody?" she wonders (Guyer 1990).

In *my* Plato, Socrates as always warns of the pain that accompanies the "dazzle and glitter of the light," as a seeker of truth is forced to turn from "the shadows he formerly saw." Yet this reading is not to settle for the conventionally virtuous belief that we must by whatever means turn from the shadows to the light. Instead, it looks to transformations in the forms and contents of discourse, "an art of bringing about."

In passages that in his Plato immediately follow the cave story, Socrates exhorts Glaucon, and us, to avoid the conventional view. "Education," Socrates says in Paul Shorey's lovely translation, "is not in reality what some people proclaim it to be in their professions. What they aver is that they can put true knowledge into a soul that does not possess it, as if they were inserting vision into blind eyes":

> There might be an art, an art of the speediest and most effective conversion of the soul, not an art of producing vision in it, but on the assumption that it possesses vision but does not rightly direct it and does not look where it should, an art of bringing this about. (476)

How shall we look where we should? Imagine a city of text. Who will rule it? What will we call it? In what way shall we write it? How shall we read it?

New Teaching: Toward a Pedagogy for a New Cosmology

We face a new world when we teach. There is no news here, for it has been ever so. Despite what we have thought of ourselves or our students, they remake us as we remake them, in reciprocal relation: no student who is not a teacher, no teacher not a student, no morning not new, at least to someone.

Our teaching always inhabits a new world, and yet, as technology amplifies newness, we find—to paraphrase the poet Charles Olson—that it is increasingly awkward to call ourselves teachers. "This is the morning after dispersion..." says Olson (1974): "If there are no walls, there are no names... and the work of the morning is methodology. How to use yourself and on what. That is my profession. I am an archaeologist of morning." (40)

How will an archaeologist in a coming age think of the world of our electronic classrooms, in which the most real objects and places are weightless and made of light, as virtual and transparent as morning air? Ann Green's and my developmental writers at Jackson Community College, e.g., work in the linked or interwoven boxes of Storyspace (Bolter et al. 1990), lifting up sections of text that others in their groups may have commented upon hypertextually and carrying these texts off in the unseen background of the virtual scrapbook, an invisible abyss somehow understood as lying upon a desktop littered with crudely drawn icons (both desk and icon constructed of square dots at the deeper resolution of "fat bits," which the students are aware of and have worked in but cannot now see). Off to the right of this imaginary desktop there floats an iconic island, not unlike that of the lotus sutra; it is the invisible, virtual space of the unseen file server, and they bump against this thing until it opens up to them like the desert of a thousand nights, disclosing yet another desktop and a series of chinese boxes in the form of folders. The students make their way down into this labyrinth confidently, through layers labeled by classes, by sections, and then by names, finally locating a folder perhaps called "Phone 1" and then choosing from a magically appearing menu another rendering of another distant, virtual machine, in this case "Modem 3."

Again they double-click, and a virtual phone rings in the speaker of the machine before them, and after some time and many more traversals, they are even further into virtuality, on an IBM 3090-600E, some 45 miles away, to which they announce themselves in ritual names as *user 2gux, user 2gut, user tfsg,* and so on, and once admitted they type an open sesame and proceed to a Conference in the form of a list of items and responses, a flattened hypertext, where they rendezvous with peers in Bill Condon's class at the University of Michigan, to whom they talk about the writings that only now do they confidently remember to retrieve and upload from the unseen and nearly forgotten virtual scrapbook where they had pasted it, long ago in both the paragraph above and Storyspace alike. After they have done all this, they linger to joke and argue, flirt and dispute among themselves, engaging in written conversations entered on alternating days (Monday and Wednesday JCC, Tuesday and Thursday UM). Meanwhile, sometimes on silent screens, televised images of Daffy Duck or leatherclad metalheads flicker, and Bach or Sinéad seeps from the speakers of the CD.

Recall that these are developmental writing students, said to be uncomfortable with hierarchical thought, unaccustomed to abstraction, unconscious of the requirements of audience. More and more I find that it is I who am uncomfortable with the requirements, not of audience but, rather, of confluence, as we push the electronic classroom to the edges of at least my tolerance for change and multiplicity.

This jumble of Bach and Sinead brings Glenn Gould to mind, who "discovered that [he] could learn Schoenberg's difficult piano score, Opus 23, if he listened to . . . [two radios], the one FM to hear music and the AM to hear the news," while he did so and who regularly relied on "the effect of . . . placing some totally contrary noises as close to the instrument as [possible] . . . it [didn't] matter what noise really" in order to attune "the inner ear of the imagination" (Friedrich 1989, 17).

The flickering moon screens, the honeycombed icons and Daffy Duck leather metalheads, and most of all their own reflections suggest, however, that our students are more apt to think of novelist William Gibson's (1987) "matrix," in which "towers and fields of [data] ranged in the colorless non-space of ... the electronic consensual-hallucination": "Bright primaries, impossibly bright in that transparent void, linked by countless horizontals in nursery blues and pinks ..." (170, 178).

I am the father of a teenager. Usually, I make them turn the sound off.

For I am, for the present at least, at my limit, i.e., seeing change, as they say on the "nets," FTF, face to face. And FTF with this impossibly bright transparent void I am sometimes unable to cope, finding myself driven to Dante to find an adequate image of all this. Even so, what might at first seem a hell, increasingly discloses itself as paradise, the place where, according to Beatrice, "All things whatsoever have order among themselves, and ... here the higher creatures see the impress of the Eternal Excellence, which is the end for which that system itself is made."

We have been talking so long about a new age, a technological age, an information age, that we are apt to forget that it is we who fashion it, *we* who discover and recover it, *we* who shape it, *we* who literally give it form. How we use ourselves and on what is how we understand the order among things themselves and the end for which our systems are made. What has changed for us, as teachers and learners, is how we see the world we remake each morning. Because we face a new world when we teach, ours must be a pedagogy for a new cosmology, a new teaching.

In shaping for ourselves, we ourselves are shaped. This is the reciprocal relationship. It is likewise the elemental insight of the fractal geometry—that each contour is itself an expression of itself in finer grain. So too, every educational institution is contoured in reciprocal relationship by the contours of each learner and teacher. Hypertext environments and networked collaborations change in fine grain the contour of learning itself, and we no longer have the luxury of thinking of computers as mere tools in our classrooms. A pedagogy for a new cosmology requires us each—learner, teacher, administrator—to choose from moment to moment among roles in this reciprocal process of shaping what Jerome Bruner has termed the forum of our culture.

Sometimes change is more comfortable if we can adapt old terms for new things, old roles for new ways. Thus, in understanding the way electronic texts shape possible contours for the culture we teach within, three well-known roles suggest themselves as apt for the new ways we are called upon to teach: scholar, teacher, and communicator. I want to briefly trace how these simple pedagogical roles shape and are shaped by the new cosmology but only after first noting a few cautionary admonitions about such tracery.

The first admonition is that no one person need take on all these reshaped roles simultaneously, or, if she does do so, she need not (and in fact almost surely cannot) take on each role equally or for any required duration. What follows from this is that an educational institution must invite these kinds of role shifts and learn somehow to take account of their productivity and encourage their changing contours.

The second admonition is to recognize that at present we understand these roles better in isolation than in interrelationship. What follows from this is that our educational institutions must avoid the easy temptation of privileging only what is already understood and thus encouraging (not to say, reinforcing) isolation.

The third and final admonition, offered with tongue partly in cheek, comes from Marshall McLuhan's (1988) only recently published *Laws of Media,* wherein he warns that we should beware of things that come in threes—that it will only be when we see four possibilities that a revolution in thinking has arrived.

There seems little doubt that technology reshapes the role of scholar. By *scholar* I mean what we know as the discipline specialist, prefaced here by the parenthetical but increasingly critical prefix, *multi.* Without going into too much detail here, I want to suggest that the role of the unidisciplinary specialist is in many ways uniquely tied to print culture and thus imperiled in this "late age of print." Or to put it in a story: at a recent Modern Language Association meeting (MLA) I shared a shuttle bus ride with a fellow who knew of my work in hypertext. He talked about how frightening this meeting was. "It used to be," he said, "that you took your degree and then you were the Victorian man, or the Restoration man, or the Shakespearean and that was that. But then you come to a meeting like this, and there are thousands of them, men and women, walking about the lobbies, eat-ing in the restaurants. And then there are people like you, hooking it all together, mixing it all up...."

Let us say, then, that in the new cosmology learning and teaching are both decentered and distributed, i.e., hooked together and mixed up. Thus, when Martha Petry and her American literature students at Jackson Community College build a Storyspace around the poetry of Walt Whitman there is a natural confluence and linkage among the machine-based learning conversation, the textual encounter, and the gathering of scholarly resources. It is a linkage that software like Storyspace is uniquely suited both to enact and to represent. The learning conversations embodied in students' journal responses not only are graphically linked to Whitman's lines as part of the encounter with the text but, indeed (and quite naturally), also find their way into the resources that the teacher-scholar brings to bear upon the text. The learners truly take their place as co-equals in an interpretive community.

The teacher as this kind of multidiscipline specialist has the important role of constructing an actual culture with her students. For we have become not merely the chroniclers or custodians of, but collaborators in, a vast cultural shift. As scholars in the dawning age of what Jay Bolter calls writing the mind, we are charged at our own peril to undertake a construction for which, fortunately, we are uniquely suited: what the poet Charles Olson calls "the discharge of the many (the multiple) by the one" (1974, 39).

The discharge of the many by the one is exactly what I mean by the role of teacher as learning manager. In an age that glorifies the business class, it is important here to footnote this usage of *management,* redeeming it in its household sense of preservation, rather than in the perverted state of channeling or consuming human resources. *Management,* rightly understood, is "making do," with each verb strong: *making do=constructive action.*

Thus, learning management is colearning, a constructive action to preserve what is coming to be known. In such a process hierarchy truly is doomed because it is before anything else a search structure, a way to keep track of what you think you own, and not an arena for reciprocal power like a hypertext, a computer network, or a collaborative classroom. Hierarchy is a model of distributed consumption, while the classroom as network is a model of distributed construction and preservation—the discharge of the many.

In *Actual Minds, Possible Worlds* Jerome Bruner (1986) characterizes "traditions of pedagogy that derive from another time... that looked at the process of education as a *transmission* of knowledge and values *by* those who knew more *to* those who knew less and knew it less expertly" (122). These "traditions of... another time" depend upon precisely the dichotomous, essentially hierarchical, categorizing kind of duality that Olson declares a dead duck. These dualities assure their own overturning insofar as they create margins and margins speak.

Over against them (by way of self-conscious and intentional duality) we can place the kinds of generative oppositions, polarities, dialectical tensions, etc., that are the very stuff of teaching, whether new or old. The classroom as network is a technology attuned to these generative oppositions and poised to broadcast and regenerate them self-similarly. These dualities likewise assure their own overturning, but by design, since they are part of a process of what Carolyn Guyer and Martha Petry call "deconstruction of priority" (1991, 82).

The teacher as learning manager *knows* that margins do not sit still and listen. She knows that margin makes meaning, and she attempts to sustain as many margins, as many fractal edges, as learners can generate. Education along these edges is not transmission of knowledge and values but, rather, reforming knowledge and values. Re-forming is re-minding. In her "notes from the margin," Cindy Selfe (1989) reminds us that teaching is a "reformist vision." Writing the mind, like any writing, is a process, and a change of mind changes how we think of ourselves in time.

The nature of mind must not be fixed. It is not a transmission but a conversation we must keep open. "If structure is [thought to be identical] with the mechanisms of the mind," says Umberto Eco, "then historical knowledge is no longer possible" (1989, 232). We redeem history when we put structure under question in the ways that narrative, hypertext, and teaching each do in their essence. Narrative is the series of individual questions that marginalize accepted order and thus enact history. Hypertext links are no less than the trace of such questions, a conversation with structure. So too, the networked classroom is a place of "making do" as constructive action. All three—narrative, hypertext, and classroom—are authentically concerned with consciousness rather than information, with creating knowledge rather than the mere ordering or inventory of the known. The value produced by the readers of hypertexts or by our colearners is constrained by systems that refuse them the centrality of their authorship. What is at risk is both mind and history.

It is an irony that in the classroom we can with certainty communicate only history. I mean history in the sense I've used above, as structure under question. I.e., we are only able to communicate the experience of a way to learn. We cannot communicate what was learned itself nor the present-tense moment of learning. The fact is that we often do not know and surely have no certain ways to measure or to judge either the outcomes of learning or its moment. Nor can we look to technology to accomplish what we cannot in this instance. Technology by its nature embodies a denial of continuity, an escape from history, in its ability to each time make do anew. Thus, it is more than ever left to the new teacher in her role as communicator to provide continuity. We are left with what we can pass on: a way of doing things, history embodied in technology.

This has ever been the case, but it is only in this self-consciously technological age that we see our way of doing things and our history so clearly embodied in machines so expensive that we hardly believe ourselves when we call them tools. In fact, it has always been the case that our tools are scarce and our history abundant. Even the simplest educational technology—teacher and students talking together—is often beyond the reach of most of the society around us. Despite our commitment to lifelong learning, very few of our students are able when they leave us to maintain the luxury of talking in community. Instead, we communicate ways to virtualize this kind of talking through the inner dialogue of reading, the polylogue of information gathering. In short, we make learning a way rather than a thing.

Ironically, the inherent scarcity underlying educational technology expands rather than concentrates intellectual capital. Forced as we are now by circumstance to relocate the locus of information and knowledge from the object to the hour—to a way rather than a thing—we re-externalize ideas and make them continuous rather than discrete, reciprocal rather than empowered, contextual rather than hierarchical. Value once again becomes embedded in the social fabric rather than the product for consumption.

Though we increasingly recognize that the classroom is a technology itself embedded in history, and that within it value is something we must renegotiate, our recognition nonetheless comes as a shock and surprise. We, too, must at each new moment learn to learn. Thus, when my student Bill Lindberg offered me his creative writing course journal as a hypertext, despite its design and links, I read through it at first as in any other term I would have read a journal in a spiral-bound notebook. It was only late in my reading that I realized that being late in the reading of this "new" document opened rather than exhausted its meaning.

A chance recognition of a simple repetition of the word *dance* caused me to understand that I had read straight through a number of screens without noticing the most obvious and richest way they made meaning. Bill had anthologized his writings from this and a previous term and then linked associatively across their texture, linking portions of screens in much the same way someone would dog-ear and underline pages to sculpt meaning out of the linear flow of a book. He had used the technology literally to highlight his thematic reconsideration of this anthology of writer's meditations and journal entries. As each new screen opened to some midway point, I would, with more or less irritation, scroll to the "beginning" of the selection and read through looking for something new. Each time, like Alexander ignorantly slashing through the knot language of the Gordian knot, equally imperially I missed the point of each midway point, the utterly evident and repeated mentions of "self," "other," "alone," and "dance," which, when I finally attuned myself to them, yielded a new text.

Even when I did recognize this new text, I was no more able to communicate in reciprocal fashion with Bill than I had been with another creative writing student, Ron Davis. I duly commented upon Ron's richly networked course journal, doing so box for box in Storyspace, until I found myself and my notes boxed and out of bounds, clearly an appendage on his visually rich webs. I should have been able to carve my own thematic response to Bill's screens; I should have been able to weave my network into Ron's. I was unable to integrate my comments within the flow of their interactions in a way or a voice that suited either student's new ways of seeing. Thankfully, at least, the technology left us all equally able to recognize what I had failed to learn from them and thus offered us each a forum in which to begin to teach ourselves how to communicate anew.

Bruner argues that "culture is constantly in the process of being recreated as it is interpreted and renegotiated by its members" and, thus, that "culture is as much a forum for negotiating and re-negotiating meaning and for explicating action as it is a set of rules or specifications for action." He identifies a number of "specialized institutions or occasions for intensifying this 'forum-like' feature," listing storytelling, theater, and law among them. "Education is (or should be)," he says "one of the principal forums," adding in marvelously deliberate understatement that "it is often timid in doing so" (1986, 123).

Bruner uses the image of the forum because it is an active public space, a place to take "an active role as participants rather than as performing spectators." I like to map the oral metaphor of the forum upon the topographic metaphor of the city of text. The city of text describes the new public space of the information age, including the electronic text itself. For the electronic text is always a forum even in isolation, teeming with multiple voices, surprising vistas, exotic sounds, or the possibility of them. Within this public space and by our actions we write ourselves upon our institutions in fractal contour. It has always been natural that talk about teaching turns to talk about writing, but now it is nearly incumbent. We live and learn increasingly within a city of text. The decentered and distributed classroom itself becomes a text that we access across both space and time. Hooked up and mixed up in a new cosmology, the classroom becomes a city of text in all its contours and white noise. Within it we must make do.

We live in a world increasingly exhausted by its own making. In our case we must remake our world continually: day to day, hour to hour, second by second. FTF with such abrupt and constant change, we literally do not know what to make of it. "What seems more compelling," quoting the poet and theorist Charles Bernstein, "is to understand (be troubled by) the situational dynamics of characterization" (1986a, 385).

In the midst of the constant murmur of the city of text we see ourselves in troubled silence. By his parenthetical "troubled" Bernstein literally means a perturbation, as in the troubled surface of shook water from which the light leaps and dances, diffuse and breathing, frangible and glancing. In such a light our silence suddenly seems not mute but musical, a measure of intervals of change. We must be troubled to see ourselves as multiple, or, as Bernstein suggests:

> to create a music of contrasting characterizations, so that you can have not only this monoplanar or dyadic movement to characterization, framing the frame, but that you can have lots of different angles in composition so that the whole sounding of the various characterizations gets heard and made palpable. (1986b, 446)

We need to see ourselves at depth and engaged: within the historical scene, not confronting it, authoring the text of our future, projecting and not projected upon. We project ourselves upon the historyless surface of technology and in the process construct the city of text. Technologies like hypertext enable this kind of "whole sounding." In constructive hypertexts we are able to see our thought in movement, to see ourselves as light and not in shadows.

Bruner (following Feldman and Wertsch 1976) calls this aspect of the language of learning "stance markings . . . invitations to the use of thought reflection, elaboration, fantasy" as a part of "culture making" (1986, 127). In a poem Olson characterizes this kind of knowing in terms of the dancer who can envision her own back as she moves in space. Elsewhere he imagines it as a slug of type, the object that knows its own form mirrored and reversed. At the time he writes he cannot yet imagine the electronic character that knows itself in each bit of arrayed, pixillated light form momentarily etched upon the phosphor of the computer screen.

We face a new world when we teach. Like the electronic character on the screen, we too must form ourselves anew. The new cosmology is contoured in reciprocal relationship by the contours of we who move within it, as participants rather than as performing spectators, as colearners and teachers. In such a tracery we will make new as we make do and truly come to know ourselves in light form.

INTERSTITIAL

Silicon Valley Maoists and Ohio Zen

Plain Talk about Technology and Education (1990)

American executives are the last Maoists. They like to have their knowledge bound up into tiny slogans written in the late-twentieth-century poetic form called the "executive summary." The person who asked for the following executive summary was an earnest and intelligent Asian American woman, not a Maoist, who did the real work for male vice presidents of the friendly computer company that kept her as a middle manager. She truly wanted to know what to tell people who asked why they should use computers in class; she wanted answers and not questions. She liked these answers and planned to use them in advertising and public relations.

For me this piece is like a roadside flare, homely, reliable, and long-burning and thus not separable from the wiry white fires and sputtering of the more theoretical sparklers. In any case the single (bottom) line still seems to emit the shower of sparks of a moral, political, and economical dilemma. Like any court poetry, the executive summary often requires command performance. For this one they flew me cross-country first-class and paid my expenses instantly. Months later the woman called to say, sadly, that most of the claims had been chopped up in the PR, but she hoped I would recognize my effect on their thinking there. I wished her the same. This year they laid her off as part of a work-force reduction and streamlining that industry analysts and the vice presidents (who, of course, remained employed) applauded.

Claims about computers in classrooms (1–3) are followed by claims about the institutional effects of integrating computers (4–6) and a last, unnumbered bottom-line claim:

1. You think differently about your teaching when you use computers, and your students think differently about their learning. Computers, unlike slide projectors or TV, offer humans visible evidence of their potential to control, change, or expand the information they confront. This advantage remains whether the learning situation or available technology allows them to make use of that potential or not and, likewise, whether they work independently, under the guidance and direction of an instructor, or in collaborative groups. The advantage grows as you increase your involvement in computing in the classroom and begin to work with networks, hypertext, hypermedia, and so on.

2. You can constantly repurpose materials you develop, taking control of year-to-year evolution of your courses and at the same time growing in your computer expertise. Class notes easily become overheads, handouts, quizzes, simulations, demonstrations, and so on. What's more, you can add to them and update them easily from the beginning, and enrich their complexity as you become more comfortable with different technologies. This interconnection is the essence of hypertext systems. If you wish to, and your institution is willing and able to support you, you can get out from under an avalanche of papers and swim through electronic alphabet soup instead.

3. You and your students already possess the major intellectual skills required to use computers within your discipline. If you hold an advanced degree, write and/or draw on the board or on handouts, and keep track of your students grades, you have the skills you need for nearly all educational computing. Computers are used in education almost exclusively for symbolic communication, primarily through word processing but also through mathematical computation as well as visual mapping and manipulation of ideas. The transition from word processing to research, discovery, invention, critical thinking, and exposition is seamless in a computer classroom, and it too is the essence of hypertext systems.

4. Educational technologists evolve in their expertise according to what they know about learning and teaching.

5. Institutions increase computer use for learning when teachers evolve in these ways.

6. Teaching and learning is more rewarding, adaptive to change, and attractive to students in computer classrooms (period).

Bottom line: What's important is that computers have a very real, almost necessary, place in meeting the needs of certain learners whose ways of thinking they seem able to represent.

Strange Talk about Technology and Narrative (1990)

These remarks were prepared for the "Storming the Reality Studio" panel (moderated by Larry McCaffery, "refereed" by Robert Coover, and featuring Kathy Acker, Ginevra Bompiani, Marc Chenetier, Takayuki Tatsumi, Samuel R. Delany, and me), as part of Unspeakable Practices II, Festival of Vanguard Narrative, Brown University 24–27 February 1993, and were reprinted in the first issue of the internet literary journal, RIFT/T :AN ELECTRONIC SPACE FOR NEW POETRY, PROSE, AND POETICS, (e-poetry@ubvm). Panelists were supposed to offer provocative and polemical tidbits about the vanguard. Most of the panel wasn't sure there was a vanguard and anyway couldn't hear one another speak and therefore didn't have much to go on. The things I remember best are: (1) Takayuki Tatsumi tapping his feet hypnotically as he read a text in memory of Kobo Abe, whose every third word seemed to be Kobo Abe; (2) Chip Delany making real sense into poetry and then excusing himself from the stage to catch a bus; and (3) Kathy Acker leaning back behind the panel table during the discussion and repeatedly whispering something that across the stage sounded like "Is sex sin?" but that later she explained was "Do you sit zen?"—which is something she does. This, too, as the executive summary above has it, is the essence of hypertext systems.

Years ago I sometimes walked through honeysuckle hills of southern Ohio with my father-in-law, a man who knew all the names of trees and the Indian cure for asthma. Because it was the 1970s, I, of course, did as my electrodes told me and romanticized his sense of nature. One time, just as we emerged from a steamy trail of maples and low sassafras into the light, he said, "Lookie there, how beautiful that is!" I tried to see what he saw against the far hills but saw only the contrail scars of a distant jet across the sky.

"What's that, Pete?" I asked.

"Them contrails," he said, "are beautiful."

"I don't know much about algebra, don't know what a slide-rule is fo'," don't know much about reality studio, but I do know, or think I know, that to be a vanguard in an age that bottles vanguard like papaya salsa will likely involve Ohio-mindedness, a constant state of oscillation and contrariness whose Zen becomes the truism etched by chase lights into terrazzo by that Ohio Zen sensei Jenny Holzer: "At times inactivity is preferable to mindless functioning."

What I'd like to propose is an Ohio Zen of sit-calm rather than sitcom. A swooning Zen in which the first thing a vanguard ought to do is give up, stay where we are, even go backward. Look for someone to surrender to; insist that someone's in charge. Which is to say, look for an edge, the temporary autonomous zone, the interstitial.

1. Forget toys. Those of us who grew up when there used to be New Cars know that technology is three-card monte and sells the future as a hedge against unhappiness about the cards on the table. The Ohio-minded shift through the whole range of Hydromatique before moving to StratoCruiser. Which is to say, make art in the technology possessed by the most of whom you think your current audience is; it will make them happy at first but then gradually perplexed about the need for a trademarked future.

2. Be like Eve: point out how they always change the names, eat the centerpiece, get dizzy at the big dance and fall down. In our garden of endless representation, aflow in the heavy water of the aleatory convergence, we think we are aware of the merging of somethings into Something: Hypermedia, multiple fiction, virtual reality, autopoesis, semantic space, simstim, cyberpunk, cyborg, grunge, rave, wax, or the discovery of television among the bees: Dave moves into the Ed Sullivan studio. "Give it up," as Arsenio says.

3. Wonka not Disney. "So much time, so little to do," instead of "small world afterall." Let our desire be a criticism that lapses before the form and so won't let form return to transparency; a criticism in which—rather than standing still—"*With each step,*" as Laurie Anderson says, "*you fall forward slightly. / And then catch yourself from falling.*"

4. Stop procreating with yourself. Whether you live in a MUD or a VR, Multi User Dimension or Virtual Reality, don't measure interac-

tion as first-personhood. Interaction manifests itself through recognition, sympathy, and witness as much as through impersonation, perception, and exploration.

5. Mind your manners. When authorship is proffered, refuse it; when authorship is generalized, claim its particularity. If they say you're an author, refuse to be; if they say everyone's an author, tell them your name is Willa or Edna. Constant declination continually renders control meaningless. We need to be content and in so being become the content of our own passionate technology.

6. Sometimes a vanguard ought to look like an old guard. Tell stories about fathers-in-law and when there used to be automobiles, have great expectations, let the dead come back to life on the overleaf. "Don't seek to master," as Cixous says, "rather... make things loved by making them known."

Part 3: Contours

Hypertext Poetics

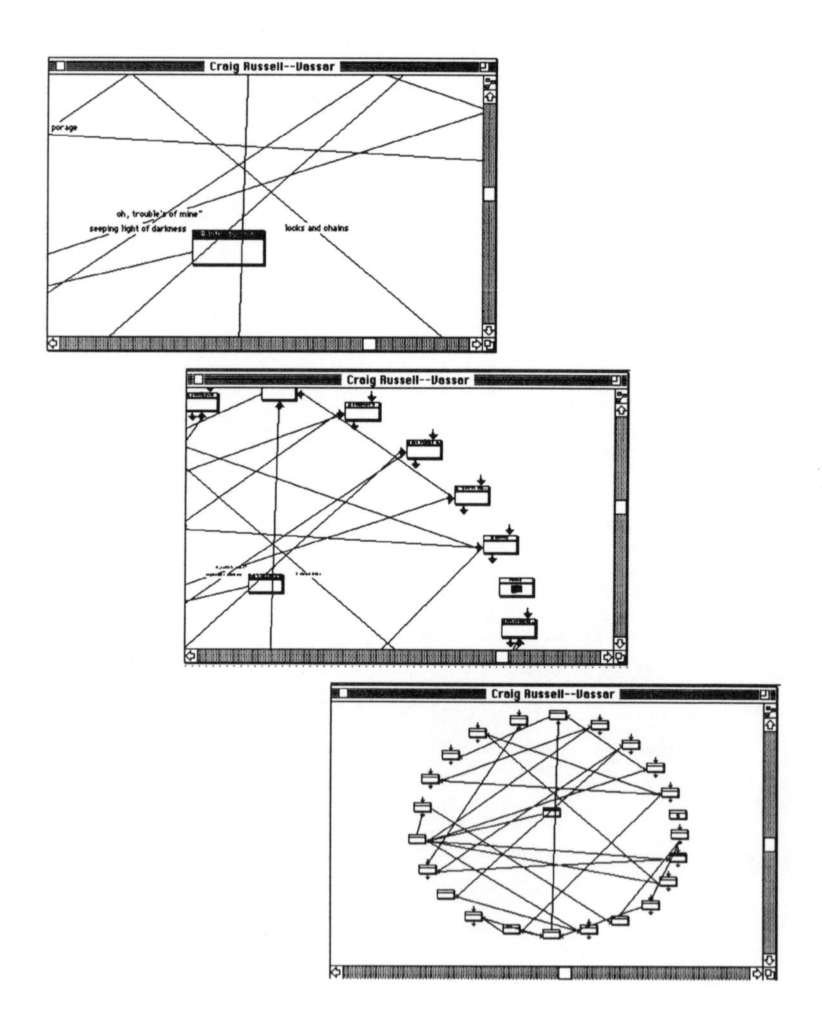

From a heliotropic reading of Erin Mouré and Lucille Clifton's poems in Storyspace by Craig Russell, Vassar '96

Selfish Interaction: Subversive Texts and the Multiple Novel

1. Interactive fictions are largely figments of our imaginations. I am not sure what to make of this sentence. It has aspects of an unrealized pun, a bad koan, a polemical claim, and a manifesto. I want to doubt it, but I believe it utterly. Though we can point to an Infocom game here and a Source serial there, when we are honest with ourselves we know that no truly interactive fictions exist.[1]

Partly, of course, this is because, among the things we are likely to think sloppily about, interactivity ranks right next to expert systems and natural language parsers. There are grounds for arguing that no truly interactive system of any sort exists—except perhaps implanted pacemakers and defibrillators—since true interaction implies that the user responds to the system at least as often as the system responds to the user and, more important, that initiatives taken by either user or system alter the behavior of the other. Video games from the glory days of the Atari renaissance or Flight Simulator and its clones or certain operating systems offer us at least a glimpse of what interaction might be. Yet try as we may to believe they truly interact, we see them branching off below us at some transparent level of the successive planes of software and hardware. We know, as certainly as we know Eliza (or her erstwhile offspring Mom and Murray),[2] that someone has been there before us.

And so we try to outguess her. Ironically, however, in this process of attempting to accommodate our thinking in response to the demands of an application's control and data structures, true interactivity does, of course, exist. In this sense text editors or databases could be said to be among the most successful interactive fictions, especially

---

This essay was published in The McGraw-Hill Handbook of Hypertext and Hypermedia, ed. Emily Beck and Joseph Devlin (New York: McGraw-Hill, 1991.

during the early stages of the learning curve as we come to use them. For during that time we convince ourselves that we know the story of our own thought at least as often as the application reminds us that we do not know its representation. Likewise, we imagine we give structure to a formless conceptual space, only to discover that the space itself is a labyrinth of glass walls within which we unravel skeins of our thought in order to find our way. An error message or a dialogue box at such times becomes an utterance from an offstage demon. We accommodate our thought to the system, and the system accommodates our thought; we interact.

This essay is an attempt to explore, along these lines, the forms that true interactive fictions might take in coming years. It proceeds from nearly three years of research (with Jay David Bolter) into developing Storyspace,[3] a structure editor for creating interactive fictions, but it is also the result of sixteen years of writing traditional novels, which were nonetheless imagined as interactive fictions without the benefit of either appropriate tools to create them or a system to present them. My suggestion is that future interactive fictions, in order to be more open, will appear more closed, i.e., more like current printed fiction, than the computer programs we currently consider interactive. The model here is, of course, Umberto Eco's, wherein "an open text, however 'open' it be, cannot afford whatever interpretation": "An open text outlines a 'closed' project of its Model Reader as a component of its structural strategy."[4] As a consequence, I also argue that we are more likely to experience satisfying interactions through the play of mind than through playing within texts, no matter how theoretically compelling the latter activity may seem.

2. *The first level of interaction precedes the creation of any text.* We live in a time when we are able to assemble more information than at any previous age in history. Even so, we sometimes seem to live as much in fear of fragmentation as in hope for coherence. Inquiry into how we process and transfer our knowledge to create coherent visions of ourselves and our worlds assumes increasing importance and is enhanced by our growing awareness that media themselves intertwine and interact and threaten to become more aether than pathway, more chaos than cosmos.

We are able to know so many things that we are uncertain how to use what we know. We have tools, but we are uncertain what tasks to put them to, since tools by their nature alter our vision of the tasks. This process is often recursive: tool alters tasks alters tool... and so on.

For a fiction writer the dynamic relationship between tools and tasks is a familiar reality. In fact, the most compelling aspect of computer tools is that they promise fiction writers a means to resurrect and entertain multiplicities that print-bound creation models have taught them to suppress or finesse.[5] I.e., where *Finnegan's Wake, Hopscotch,* or, for that matter, *Tristram Shandy* created multiplicities as intricate as any of those we envision for interactive fictions, those texts were bound at the very least by the static nature of their presentational systems.

As a result, Sterne, Joyce, and Cortázar each create unique, and illustrative, solutions to the static, linear presentational models open to them. Since metaphors and metalanguages alike inform through juxtaposition, it might be useful to consider these authors' solutions in terms of what—by Joycean "liesense"—we might call "computertease."

Sterne exploits the decorative and self-referential qualities of the Gutenbergian medium in what we might call a "screen-based" mode. Like John Barth after him, or the medieval copyists before him, he recognizes that the graphical coherencies and conventions of the printed page can be counterposed to the textual linearity. In this sense Sterne anticipates graphical user interfaces, making a book about part of a book look like a book of parts of books.

James Joyce attenuates language in what we might call the "line editor" mode, recognizing that imprinting is *in*-printing and thus eidetic. Like the Lewis Carroll of "Jabberwocky," he forces a "what you see is what you get" environment to yield what we might call virtual windowing, in which the in-printing seems to expand and disclose a momentary flicker of other words, other languages. His process thus is also graphical, allied as much in his time with collage or Ezra Pound's notion of ideograms as with Knuth's metafont or Macintosh's Cairo font in our own time.

Joyce, however, compounds and parallels these eidetic qualities of print with narrative "macros," which summon "overlays" from what we could call—quite rightly, in Joyce's case—"libraries." These interlace successive text segments and cause "side effects," which extend the flicker of a line to a cyclic pulse of whole pages.

Whereas the effects of Sterne and Joyce reside largely in the record fields themselves, the effects of Cortázar are strictly relational. *Hopscotch*[6] comes with what we might call "subroutines" (or perhaps "shell scripts") that are not at all unlike Infocom branching structures or childrens' "Choose Your Own Adventure" books. Whereas Joyce and Sterne overtly reference a system library of parallel texts and thus make their language what we might call "declarative," Cortázar provides a procedural language that readers may use to reference their own libraries.

I.e., he calls upon us to recall not the subject but, instead, the act of our previous readings and invites us to read *Hopscotch* in a successive, literally programmed iteration. This second reading depends upon our decaying, yet dynamic, memory of the master text.

Whatever their differences, each of these three novelists surely pre-processed interactions with their readers, anticipating in their input stream, so to speak, the peculiar limitations of the batch-processed, linear output stream of a static presentational text. While this kind of pre-processing is, jargon aside, surely an aspect of any novelist's creation of an ideal reader, these novels are distinctive to the degree that their texts overtly confirm "correct" traversals.

In other words, the pre-processing results in the creation of an intricately networked novel-as-knowledge-structure that both simultaneously invites and confirms reader interaction.[7] Moreover, the text itself carries a syntax for such interaction (an implied labeling of network arcs) that depends upon a reader's familiarity with the operating system of literacy. I.e., we readers are invited to confront these novels, even on first reading, in the same fashion that we have previously flipped through, browsed, reviewed, or recollected other, more linear novels.

This invitation to confront is what I believe instantiates the interactivity of these texts, offering, as it were, the "structural strategy" of Eco's open text. Each of these novels is what Eco calls "work in movement." As such, "the possibilities which the work's openness makes available always work within a given field of relations": "We can say that the *work in movement* is the possibility of numerous personal interventions, but it is not an amorphous invitation to indiscriminate participation. The invitation offers the performer the chance of oriented insertion into something which always remains the world intended by the author."[8]

Therefore, it is perhaps not too much to say that, if truly interactive computer fictions are to exist, they will require tools for simultaneously creating and recovering such oriented insertions within given fields of relations. I.e., interactive fictions require text-processing rather than word-processing tools. For text processing, unlike word processing, is a method of intellectual and artistic discovery and presentation. If word processing may be thought of as a tool for thought (making any written text available for emendation, elaboration, and restructuring at any point along its length), text processing may be thought of as a thoughtful tool (making a master text and variations of it available to readers along a presentational path predetermined by the writer and selected or influenced by the reader). Text processing makes texts transparent, inviting readers to consider parallels, explore multiple alternate possibilities, and participate in the uncertain process of discovery and creation.

Our program, Storyspace, originated as an attempt to develop a text-processing tool to enable writers of interactive presentations to exploit multiplicities. Storyspace depends upon a decisional order rather than a fixed order of presenting material. A fixed order presentation may be thought of as a road map, within which the author may present side trips or forks (some of which the traveler or reader may be allowed to choose among). The order of presentation, however, remains linear. The trip (or text) proceeds from point A to point B, however various the digressions.

A decisional order, however, may be thought of as a series of locales, some of which are linked by linear progression or argument, but with others determined by allusiveness, resemblance, evocation, or unexplained and "intuitive" parallels determined as often by the author as by the reader. The journey along a decisional pathway may continue to proceed from point A to point B, but the traveler will not only be invited to dream or recall other journeys (some as yet unmade), but also will be confronted with unexpected or surprising turns, detours, and compelling alternate routes. More important, each alternate reading will cause the text itself to degrade or reform, so that no successive reading will ever again substantially parallel a traversal of the initial master text. In such a way the reader's creation of an implied text could become physically enacted on the screen or in instantaneously laser-printed (and, one could imagine, perfect-bound) hard copy.

This kind of subversive text, or multiple novel, would offer its readers an opportunity to explore the cohesion of a work in a way analogous to successive readings separated by time. The reading would, in this sense, encompass in its interaction either the kinds of imaginary dialogues we often conduct with author or character after reading a story or the reveries of succeeding (or "what if") episodes that either precede, interject, or follow.

Rather than speculating on a character's deeper motivations, e.g., the story itself might, in successive readings, offer greater or lesser explorations of motivation. As such, the text would overtly offer in its structure that which Eco claims for every work of art. It would be "effectively open to a virtually unlimited range of possible readings, each of which causes the work to acquire new vitality in terms of one particular taste, or perspective or personal *performance*."[9]

To enable potential interactions of this kind, the Storyspace text processor preserves both a decisional pathway (the choices made by a creator of a work, i.e., the interaction that precedes text) and a presentational pathway (sequences of nodes and their links available to the audience/reader, i.e., the interactions that follow from texts). The decisional pathway provides the creator of a work with a repertoire of associations and linkages, some designed solely to serve as mnemonic, architectonic, and editorial stimuli and utilities, and others designed to directly interface with the presentational pathway. This latter pathway provides a means for presenting multiple, interactive versions of a work, with the range, liberty, and syntax of interactivity determined either by the creator or by the reader. The text thus may be accessed either directly or through predetermined associative scripts designed by the author but transparent to the reader.

Storyspace might thus be used to create a novel as supple and multiple as oral narratives, but with the referential and coherent richness of in-printed, relational works such as those of Sterne, Joyce, Cortázar, and others. These computer-enhanced, subversive texts would merge process and product much in the manner of Baroque themes and variations or jazz improvisation, in which the colorations or embellishments are ephemeral, often depend on audience or occasion, and usually resist static apprehension, or capture. The intellectual basis for such work involves both a natural collation of trends in twentieth-century thought concerning the transitory and multiple nature of human experience and also a reflection of our widespread cultural urges and individual longings for methods to identify and represent the perceived order and complexity that underlies a mass of information.

To the extent that the decisional pathway would also be available for readers to use, we might also expect that they would template story-generation "scripts," which would offer them opportunities to participate in the narrative itself in the fashion of a hypertext. Certainly, the narrative and the technological models exist for readers to do so.

Yet would they? I think not. For a fiction is essentially a selfish interaction for both its author and its reader.

3. It is likely that no one interrupted Homer. Or, perhaps there were drunken hecklers even then, in the courtly supper clubs of Asia Minor. Almost certainly there were the Attic equivalents of piano bar patrons who politely request "Melancholy Baby" or "Steadfast Penelope." Even so, we can feel reasonably certain that no outsiders interrupted Homer in the sense that they established a priority in his narrative or that their language subsumed his.

Or to put it differently, a question we might ask ourselves in our enthusiasm about interactive fictions is: *Since the technology has existed for some time, why don't people write alternate chapters in the blank spaces of bound novels or alternate sentences in the blank spaces between printed sentences?*

Why don't most readers write at all? Why don't most readers at least write the beginnings of sequels to the novels they love?[10]

We would like to respond that the medium was not appropriate, since we would like to believe that the computer medium *is* appropriate. Yet another series of questions follows upon these proposed beliefs.

Why don't most readers of current interactive fictions write alternatives within them or sequels without them? Certainly, the medium offers opportunities. Users of multitasking systems could input alternatives in one window as the interactive story scrolls in another. Users of other systems could summon text editors running under the application or, as an emergency expedient, use break keys and input alternate lines to an uncomprehending operating system or monitor.[11]

To a certain extent, of course, users already do these things. Even the most experienced users of interactive adventure or mystery games sometimes find themselves inputting commands that exercise the parser beyond its limits and cause it either to admit ignorance or output default, Eliza-like solecisms. And few honest computer users can swear that they have not explored the variety of canny, canned responses that applications or systems programmers provide in anticipated response to the hallowed Anglo-Saxon expletive and imperative.

The two preceding cases suggest a paradigm for what I want to call selfish interactions. In the first, readers either become too enveloped in the language of the narrative or, paradoxically, fail to become fully cooperating participants in the linguistic subset that the parser demands. I.e., they either lose themselves in the story world or experience the loss of the story lexicon in the parser. In the second, users acknowledge the linguistic subset by seeking ironic or hortatory confirmation of it. I.e., they explore the limits of phatic communication that underlie a highly codified linguistic subset.

In each case, the attendant "pleasures" are ones that involve confirmation of interaction. Yet what is confirmed are limits, or the traversal of boundaries between procedure and declaration, i.e., between the act of reading in a certain way and the text, which anticipates and signifies successful reading in that certain way.

Thus, we are unlikely to write intertextual, parallel, or detached versions of an interactive fiction because we selfishly seek confirmation of our alternative choices within the text. Likewise, we are more likely to write (or utter) alternatives to a fiction either within a computer conference, in which other readers can assume author or character roles to confirm our choices, or in adventuring or other groups, in which conversation and mutual fantasy supply confirmation outside the software or text.

Yet the communal aspects of these conference or group readings are likely to be unsatisfying to the degree that any participating role player moves the text beyond the constraints we imagine the author imagined. Or, to state the case positively, we are satisfied and delighted when the text reconfirms our variations. Thus, an interactive mystery that presents a taunting prompt, i.e., "Too bad you dropped the phaser, you could sure use it now," delights us even as we are reduced to carbon atoms. It is unsurprising that the overtly intertextual nature of most current interactive fictions is so attractive to literary theorists and narratologists. I.e., we often read interactive fictions in parallel streams consisting of (1) the narrative stream, or the story itself, which echoes, self-reflexively quotes, and anticipates its variations within a relatively unconstrained lexicon, and (2) the game stream, or the "story" of the narrative flow chart and program syntax, which not only enables confirmations within the first stream but also provides confirmations of its own, but within a highly constrained, formulaic lexicon. We are as pleased to recognize the structure of the latter as we are to experience the structure of the former.

Yet the current apparatus seems somehow makeshift. The pleasures of parallel reading are not quite compensation for the jolt of alternating lexicons. The continued development of more successful parsers cannot conceal the fact that true natural language front-ends are years away from development, nor can it convince even the most naive users that they have engaged in genuine interaction resulting in alterations to their reading behavior and the story's text. We know that we are engaged in something more like a game than a reverie, and, although there are undeniable pleasures in mapping the story and its branches while matching wits with its programmer within the game stream, we somehow know that in the narrative stream we are interrupting Homer, or, worse, both Homer and his programmer.

Ideally, of course, both the program "map" and the text it traverses would change with our changing interests, much in the way, e.g., that Joyce's *Ulysses* changes style to suit sensibility, quarter of the city, and hour of that June day.[12] I.e., it is possible to conceive of an interactive computer fiction within which the branching schemes, alternate traversals, and so on both disclosed themselves as integral parts of the narrative and altered themselves according to our shifting interests and current states. In such an interactive fiction the branching choices, attribute lists, and implicit/explicit menus would disappear into a seamless, but subversive, text, which constituted both the story and the game stream. The immediate effect would be both to widen the fictional domain, or story world, of interactive fictions and, more important, to involve a wider audience and, thus, a wider authorship.

A subversive text would restore the Homeric situation to the extent that its branching would be driven by (1) previous conditions and (2) detection of audience interests, rather than by direct intervention of the audience into the story. As an example, consider the following, purely hypothetical, interaction we envision for Storyspace.

The interactive fiction, *Emma,* consists of a paperback book (or master text) and an accompanying computer disk. On the disk are the master text and subversive variations. The screen display is very similar to the Macintosh notepad in that it allows paging and browsing forward and back in the text. Each screen "page" is identical to a portion of the corresponding page in the printed master text. Thus, a reader who preferred screens to printed pages would be able to read *Emma* through. Once.

Without browsing forward or back.

But she would have no indication of this fact, except that when the reader flipped forward, e.g., in the screen master text from page 2 to page 200, it would be impossible to return to page 2. Or, more accurately, the page 2 the reader returned to would be utterly different from the page 2 she left.

Since the master text would still exist in the printed version (and would have existed on the screen for one continuous read-through before it decayed), it would be possible to try to make some sense out of the second "page 2" by comparing its intent, focus, style, point of view, etc., against the master "page 2."

Perhaps the reader could test a hypothesis (still very much in the game stream fashion) by browsing forward again from page 3 (using either the master text as a map or using a search function within the program to choose the same character or setting that the previous browse to page 200 resulted in). Or perhaps the reader would simply read sequentially through the "new" story, choosing to compare it or not with the master text.

Suppose, however, that the reader *did* choose to pause at different points and compare with the master text. Suppose further that the author had embedded in the decisional pathway a branching structure that depended upon the machine's ability to calculate (and/or accumulate) "eventless" reading of certain pages with common themes or characters. Suppose that branching structure presented yet another version of the multiple novel, with an ordering and structure quite unlike the hypothetical one the reader has decided to pursue.

Suppose that in the very next episode a character seemed to recall an event very much like the one that used to be on page 2 and asked someone's help in helping to recall it. And the screen display suddenly refused to move to the next screen page but would move forward and back through everything that preceded the current page and followed the page after next.

Suppose the reader quickly gets frustrated, puts the disk aside, and does not use it again until months later. Perhaps winter has changed to spring in the interim. Suppose that the "spring" novel were entirely different from the "winter" novel and that the computer's clock triggered the new version.

Or suppose that each time the disk was rebooted after a master reading, a blank screen appeared, or an old-fashioned *A>* prompt or a *?* cursor or a polite request, "What would you like to read?"

Suppose that there really were natural language front-ends and expert systems that constructed reader "agents" who tailored texts to suit the input to such prompts. Or imagine a technological leap resulting in both faultless voice recognition chips and natural language processors so that random mumbles or directed beefs and bravos created such agents and the resultant new texts. Suppose that input to the above noncommittal prompts caused the software to take over the system and dial up 800 numbers, which downloaded new versions (or caused Videotex services to pump new text through the friendly TRINTEX or AT&T blackbox in the living room).

Suppose that the text *did* invite you to write a version, an alternate sentence, a chapter, a sequel? And then never showed it to you again. Or showed only parts of it, intertwined with a version of the multiple tale appearing months after the reading that prompted you to write in the first place.

Suppose somebody shouted, "Play Steadfast Penelope," and the screen display altered to Greek fonts or the voice chip began to declaim formulaics in a foreign tongue.

4. Suppose a text can anticipate unpredictable variations upon it. I am not sure what to make of this sentence. It has aspects of an unrealized pun, a bad koan, a polemical claim, an oxymoron, and a manifesto. I want to doubt it, but I believe it utterly.

There is a phenomenon known variously to computer scientists but often called *interference*, the *concurrency problem*, or *software interaction*, in which, in complex systems and programs, levels upon levels of software functioning together exhibit unanticipated "clobbers" or "side effects" causing unpredictable and usually nettlesome, but occasionally felicitous, results. For a debugger these interferences call to mind the swamp of "The Big Two-Hearted River," where "the banks were bare, the big cedars came together overhead, the sun did not come through, except in patches [and] in the fast deep water, in the half light, the fishing would be tragic."

In a similarly dark place Eco considers a question (parallel to interference or the koan that begins this section) that exhibits itself in James Joyce's prediscovery of nuclear fission in the phrase "abnihilation of the ethym," in *Finnegan's Wake*. According to Eco, "the poet anticipates a future scientific and conceptual discovery because—even if through expressive artifices, or conceptual chains set in motion to put cultural units into play and to disconnect them—he uproots them from their habitual semiotic situation."[13]

The discourse here is also a swamp (and the fishing is dangerous), yet what comes through is a vision of interactive interference not only anticipated but also, in a sense, *forced* by the networking of language. "Sooner or later someone understands in some way the reason for the connection and the necessity for a factual judgement that does not as yet exist," suggests Eco: "Then, and only then, is it shown that the course of successive contiguities, however tiresome, was traversable or that it was possible to institute certain transversals. Here is how the factual judgement, anticipated in the form of an unusual metaphor, overturns and restructures the semantic system in introducing circuits not previously in existence."

This kind of fishing, of course, exists at the level of the linguistic microcosm, yet it does not seem so far removed from the kind of macrocosmic interaction of dual story streams that I have suggested as the most likely mode for coming interactive fictions. The linking arc in such a network is the selfish reader, the someone who sooner or later understands the reason for a connection.

What is critical, however, is that the arc itself be literally instantiated, as present in the creation as in the performance.

"The factual judgment," says Eco, "draws, perceptively or intellectually, the disturbing data *from the exterior of language*": "The metaphor, on the other hand, draws the idea of possible connection *from the interior* of the circle of unlimited semiosis, even if the new connection restructures the circle itself in its structuring connections."

The multiple novel, likewise, will invite writer and reader to restructure the circle of a text in its ability to simultaneously, and subversively, present both the exterior and interior of language through successive, and shifting, story streams. Already the tools present themselves for use: the idea processors, hypertexts, and Storyspaces.

"[Nick] looked back. The river just showed through the trees. There were plenty of days coming when he could fish the swamp."

INTERSTITIAL

Dead White Men Also Compute

Can(n)on fodder/fæder: *The above means to characterize the not very subtle, subversive rhetoric of a piece that was solicited for a grant request as an argument to convince "traditional humanists" of the potential for computer pedagogy and hypertext. Herein are quoted more august white men per capita than anywhere else in my work, though to little effect, since the rhetoric was not something I was able to sustain for long. By the scenario the white men begin to dissipate like fog, and the humanists saw through it all anyway. I never saw any of the money.*

> "But this person who, we say, has only belief without knowledge may be aggrieved and challenge our statement. Is there any means of soothing his resentment and converting him gently, without telling him plainly that he is not in his right mind?"
>
> "We surely ought to try." (Plato, *Republic* 5.476)

Channels full of late-night cable television hucksters and generations of self-help books tell us that "every challenge presents an opportunity." For humanists, educators, and media professionals alike, this is a message cheapened by each telling. It is a message as infected by suspect intellectual values as it is sapped by poor production values. We would reject it out of hand, if only we were not so implicated in it.

For both the message and its media are ours. The humanities refresh and renew themselves under challenge. They transform human beings by challenging them to discover opportunity. What's more, the humanities have ever insisted that refreshment, renewal, and transformation can be accomplished by media: the simple transmission, to eye and ear, of what women and men say, write, and do as they act most humanly.

Yet how, we ask ourselves, are 13 units of computer-enhanced instruction on a CD-ROM so different from 20 self-help audiocassettes in a vinyl folder for $59.99? Nontraditional learners, seduced either by 800 numbers promising six steps to immediate wealth or three tapes to positive self-image, are not so different from traditional learners, seduced by three-year management programs with starting salaries of $800 weekly.

How can we capture their imagination and, more important, encourage transforming action, with our message of challenge and opportunity—whether Socrates' "Know thyself" or E. M. Forster's "Only connect..."? How can we revalue what has been cheapened? How can we lead others to reevaluate and discover the opportunity in current challenges?

These questions do not outline a straw man argument but, rather, a continuity of perhaps the oldest debate about humanities education and human media. At its heart is the difference between beliefs and knowledge, between semblances and essences. Late-night hucksters, some three-year management programs, and, indeed, too many educational technologies encourage mere beliefs, i.e., what Plato, in *The Republic*, called dreaming: "whether awake or asleep... mistaking the semblance for the reality."

The humanities, however, insist upon knowledge, i.e., what Plato describes as the ability to "discern... essence as well as the things that partake of its character, without ever confusing the one for the other." This is not an easy task. And yet it is a task to which new technologies are uniquely suited. Discerning essence from semblance is what any educational medium does. At its best a medium engages us through our beliefs and then calls upon us to transform beliefs to knowledge. Watching a television program about the Milky Way may lead us to believe, briefly, that we inhabit the stars. When we understand that the living room in which we view the program inhabits a planet among the actual stars, however, we possess true knowledge.

Both the semblance and the essence—the body and mind—of the humanities are under challenge. Recent research suggests that college students don't master the current body, or corpus, of material even as that corpus itself is under challenge. In our several disciplines we engage ourselves with the opportunity to reconsider that corpus. We think that reforming the body may reform the mind.

Yet we are, as always, skeptical of the mind, more so as the mind is extended by new media. Evidence suggests that college students increasingly cannot demonstrate critical reasoning skills. Yet the ability to deal with the semblance of ambiguity or to decipher complex events, issues, or statements is the essence of what we do in the humanities. The current challenge seems to transcend what the students do or do not know. Not just our perception but also the value of the humanities itself seems at risk. In our disciplines we engage ourselves with the opportunity to reconsider our processes of thought. We think that re-forming the mind may reform the body.

Whether to reform the body or the mind is a dilemma that confronts not just the humanities but our culture. As a culture, we have to contend with an explosion of information and information technologies. As humanists, we know that data is just noise if we can't decide what matters and what doesn't. Otherwise, the challenges we face become commonplace. They, too, risk becoming messages cheapened by each telling.

> We must find ways to bring history into the computer world. . . . It is a truism but nonetheless true that we are about to witness a great transfer of knowledge from one medium to another. . . . In [the place of the old medium] would be the qualities of the computer age. . . . the sense that recorded knowledge can be collected, transmitted, reorganized, and molded to one's immediate needs and the feeling of being immediately "in touch with" our culture's past. The technologist might find the past more accessible in this form, and the humanist might find new uses and interpretations for familiar texts and historical events. (Bolter 1984, 225)

Given the right size group—e.g., a small cocktail party—separate conversations tend to coalesce into one. This is not necessarily because one conversation is more important per se but, rather, because it promises to encompass and focus the others. Those on the edges of two separate conversations invariably find themselves participating in both. They link concerns and share perspectives. Like the computer game of Life or the workings of a healthy news organization or educational institution, linkages are made and new forms result that retain the best of the old forms. These conversations find at least three settings in contemporary academe.

This image of merging conversations describes the classroom, the primary setting in which, as learner-teachers, we have the opportunity of presenting the humanities as a process of shared inquiry. In this setting we quickly move beyond the mere semblance of communicating the existence of certain books, problems, or ideas. Instead, we center upon the challenges our students confront in seeing the humanities as both a model and a method. We encourage students to identify our models and evaluate our methods for approaching our shared problems as learners. Four separate areas of inquiry can easily be framed as questions in this ongoing conversation:

1. What sorts of personal involvement in books, ideas, or problems do learners experience?
2. What kinds of attention, rather than how much attention, do learners pay to what they experience from books, ideas, or problems?
3. What characterizes learning as an interpretive process beyond what a book, idea, or problem in a local and contingent way "means"?
4. How adequate are the humanities in providing a model for involvement, attention, and interpretation of what learners experience in their lives?

Classroom conversation obviously spills out into hallways and becomes conversation about the appropriateness of both classrooms themselves and the institutions that house them. Each of us in our separate institutions engages in conversations about educational institutions and change. When hallways begin to talk, whole institutions echo and, eventually, create a second setting as we hear one another. Again, common questions emerge:

1. What, beyond educational fashion, leads faculty to form interdisciplinary groups, and how can these groups contribute to refashioning an institution? What leads such groups to seek out one another?
2. What qualities characterize our students, whether "new," nontraditional, or "old," career-oriented; and how can we address both their diversity and commonality?
3. What challenges do such new/old, nontraditional/careerist students present for teaching and learning; and what new designs for teaching and learning can we devise to meet those challenges?
4. What, beyond evaluation and accreditation, should an educational institution offer a distant learner; and how, if at all, does this differ from what we ought to provide the independent learner we wish to encourage on our campuses?

5. Why are we drawn to computer learning environments; and what in hypertext/hypermedia promises to satisfy our shared vision of the future of education?

The third setting for conversation was, in fact, not much different from the first two. Even so, we term it a machine conversation. For the computer offered the focus and encompassing qualities that merged all three conversations. Although we know better, we cannot avoid the feeling that our conversations have been carried on as much with the machine as about it. We have conversed with the machine by trying, with tools like hypertext, to design a model for humanities education that addresses questions from other conversations: questions about learners, institutions, and the humanities themselves. Hypertext lets us see and manipulate these questions in a way that we, as readers, control. More important, it lets us see and test the underlying structures and associations in what others say against our own structures of thought. Like Tron, we have become implicated in the design of the information we inhabit. Unlike Tron, we seek, rather than risk, our own identity, even when we confront the thought of others.

This quality of designed implication differentiates our project from other computer-enhanced learning projects—even other hypermedia projects. Such a differentiation requires a longer list of questions:

1. Can we use hypertext to produce a new sort of "book" intended to involve readers in producing books of their own? Can we move beyond the recording of reading and decision making, which some hypermedia programs do, to let learners discover the consequences of different interpretive strategies, including our own?
2. Can we implicate these reader-learners in the design of the knowledge before them, letting them discover or propose what "works" given a certain point of view, what doesn't work, and what, ultimately, it means to have a point of view?
3. Can we create a genuinely new, reader-centered, graphically represented, method of constructive interaction and instruction for the undergraduate humanities curriculum? Can we combine in this instruction the best qualities of traditional and nontraditional teaching and learning?
4. Can we embed integrative, evaluative decision making in the program itself, rather than simply in the instructor's discourse, so that

distant learners can avail themselves of instruction as well as the syllabus? Can these same integrative, evaluative qualities serve instruction in more traditional settings, freeing teachers for inquiry and co-learning rather than exposition?

5. Can we engage reader-learners in continuous evaluation of their own experience, by showing them that they are clearly involved in and responsible for choices they make in education as well as in life?
6. Can we convey the essence of basic texts, events, or ideas by enabling reader-learners to explore our relationships with texts, events, or problems? Can we lead learners to transform this exploration of knowledge into real inquiry, and real relationships, of their own?
7. Can we lead reader-learners to discover that these activities—engaging, evaluating, and enabling—are both the natural consequence of using the computer as a mode of inquiry and also the essence of teaching and learning in the humanities?
8. Can we design this instruction so that it can exploit, but will not require, current resources in other media, such as videodisk, CD-ROM, and more traditional audiovisual media? Can we design it so as to anticipate, in methodology if not functionality, future developments in educational technology and knowledge resources, including public "information utilities" such as videotext and portable learning machines or personal libraries; or the evolution of existing libraries into omnipresent, networked, and distributed multimedia, "learning resource centers"?
9. Can we justify the enormous expense, relative unavailability, and limited duration of most students', especially nontraditional students', current access to computers? Can we provide knowledge gathering and knowledge design skills whose real, economic value offsets these expenses and maximizes the benefits of limited availability and access? Can we make these knowledge skills marketable outcomes of the learning experience, which students can use for entry into careers and professions?

We do not propose to answer all these questions or to integrate successfully these conversations. Yet we do propose to formulate our design, and evaluate our efforts, according to them. "The mind is a preliterate society," says the poet David Antin. Yet we face a radical turn toward postliteracy: topographic writing, hypertext, and a machine-based conversation of constructive association not unlike the workings of the mind.

As a first step toward a hypertext for the humanities, we might imagine something very much like hyperfiction in conception or at least in its branching, convinced that the mix of seamless, default branches, yields, and patterned browsing that characterizes them is something like the right model for interactive texts in any domain, and especially for the kinds of choice-driven modules we are discussing. The ability to design and implement an adaptive circuit logic for branches is a strength of hypertext and enables the learner to experience the sensation of a constantly changing text.

Unlike most hyperfictions, however, the humanities modules would generate a model of the students encounter with the structure of thought. At some point students should be able to compare this model with the authored structure and organizing elements of the module itself and, more ambitiously, to create an extended network that follows from this first encounter and links to other modules, some of the students' own creation.

In a scenario of sorts the student using a module will essentially encounter the title page for an electronic essay, the Spine. This title page will offer a choice between clicking for directions or continuing. The directions will be a simple introduction to hypertexts and the humanities and will suggest that the student may either click on any interesting words, phrases, or graphics (Yields) with the result that the text will change, or she may simply choose to page through the Spine using the defaults. The Spine will include a browser, which lists possible choices and directions, but none will be displayed until the student has attempted and/or successfully hit upon a Yield, the notion here being that a text only takes on dimension when the reader decides to act upon it.

The Spine text will largely have the flavor of a coda, an expository but fairly personal traversal of issues, methods, allied questions, problems, etc. If the student chooses a default path through the Spine (no attempts to hit a Yield), she will eventually reach the first node of a subset of nodes on alternate paths (which we can call Resources), which otherwise would have been accessed, often midway, by seeking a yield from the Spine.

The Resources are not types of texts or graphics but, rather, structured approaches to inquiry developed by the authors and perhaps (eventually) augmented by the student herself. Resources might be grouped along networked, interconnected paths termed, e.g., Personal Encounters, Collaborative Work, Primary Documents, Other Responses (including secondary documents), Open Questions, Prejudices, Author's Journal, Rhetorical Considerations, Linked Modules, and so on.

Resources can be visualized as convergent paths that inform the Spine, i.e., a sort of intellectual "audit trail," the actual work space of the authors themselves and eventually their readers. Resources are plugged into the Spine as part of its creation, and they remain there later as gateways to the hypertext the student will eventually create by her reading. Resources thus serve as an intellectual map of both the authors' encounter and the authors' proposed mentoring or conversational relationship. They should be gateways to a body of material that has exactly this conversational, mentoring quality but that does not attempt any of the unctuous mock interaction of Computer Aided Instruction. Instead, these nodes should have exactly the flavor of the kind of discourse we use in mentoring and advising students or giving informal lectures. This Mentor path should be a privileged resource vis-à-vis the system of the modules, i.e., one that "grows" an increasing number of links to the Spine and other Resources as the latter are explored and that eventually opens into full conversation with the students who create its network of mind.

Network culture *extends this discussion to the World Wide Web and beyond. It was first solicited as a statement for a working group as part of the National Initiative for Humanities and Arts Computing sponsored by the Getty Art History project, The American Council of Learned Societies, and the Coalition for Networked Information. This version appeared as a statement for the briefing book of "Convergence: Art, Culture, and the National Information Infrastructure" (October 1994), a conference sponsored by the Center for Art Research, The New Art Center and the MIT Artificial Intelligence Laboratory at the State House in Boston.*

In developing Infobahn policy, entertainment industry forces (what Stuart Moulthrop calls the "Military Infotainment Complex") are largely calling the shots. Thus the New York Times (September 12, 1994) reports an advisory group for a digital Library of Congress that includes The Walt Disney Company, Bellcore, and the National Science Foundation but no artists or humanists.

Artists and humanists increasingly depend upon the network as a place for performance, community, collaboration, and publication. Though we may fancy ourselves as functioning at the interstices in temporary autonomous zones, we increasingly find that our own "cultural heritage" is, in the widest sense of the term, the net itself.

Fostering the activities of artists, scholars, learners, and audiences on the network requires a locale-based, rather than a resource- or tool-based, approach to continued development of electronic communities. Infotainment/infobahn projects ought to require cultural impact statements taking into account the following factors:

Assured access. Assured access (a "non-negotiable item" in the way of the abandoned universal health care coverage) must be bidirectional and multimodal, not merely a consumer service (even an interactive one) but an extensible, collaborative environment. Bandwidth-in must equal bandwidth-out. Access must extend to both artists and audiences, to scholars, teachers, and lifelong learners, to those with both local institutional affiliation and remote or virtual affiliation, and must include a carefully conceived public space supported by librarians, scholars, teachers, and artists. The new culture needs a new generation of Carnegie Libraries. Electronic mail and the Web must be universally available since much of the development of the internet has been fueled by the low-level interactions and extensible formats that e-mail and the Web offer.

Extensible formats. Clive Smith, Geoworks:

> The expressive vocabularies available . . . are significantly contingent on whether the software (UI's, emoticons, digital ink, navigation tools, search/link facilities, etc) of the Net (or of the dominant consumer services) are *extensible*; and especially whether they are *user* extensible by publishing the extensions as embedded facilities within any given message.
>
> . . . the potential range of artistic innovation/vocabulary can be massively larger than it is if artists can mainly recombine what is already there. And the rate of diffusion for those innovations can be much much faster, because [they can be sent] to any recipient, regardless of whether the receiver has the same tools . . . or whether the facilities are standard Net protocols. Messages can always default down to the hardware limitations of the client/receiving machine. (Smith 1994)

Cultural factors design and interfaces. Human factors research considers the design of interfaces, data structures, computational objects, and so on. Likewise infotainment media depend upon focus groups and other target-marketing techniques from inception to release to after-marketing of entertainment and informational products. These commercial and technological activities only occasionally—and only marginally—enjoy the benefit of cultural factors research, of which the arts are a special case. Examples of cultural factors range from multicultural aspects of technological interfaces to the loss of cultural memory that results from decisions made by electronic record keepers. Assessment of cultural factors should become a standard component of any network service, infotainment product, or technological environment. The arts and humanities are uniquely situated to develop, broaden, and disseminate standards for judging cultural factors in technology through public discussion, publication, criticism, teaching, as well as social and political action.

Cultural Impact Statements

Recent rounds of trade talks within the EEC and long-standing Canadian social policy regarding entertainment products imported from other countries signal an awareness of the cultural impact of technology. U.S. federal policy—as well as the "shadow policy" established by private philanthropies, transnational corporations, and the media—should require that cultural impact statements become part of the licensing, regulation, and review policy of any publicly or philanthropically funded media or infotainment project, as well as part of the business plan or capital formation strategy of any privately funded project.

The Geography of the Word: The Textfile as Landscape

Although we come to know the world in time, our intuition is that we recall it in space. An instinctual, naive neurophysiology seems to confirm our intuition. A child, e.g., will sometimes press a hand against the skull in an attempt to remember. Likewise, any of us may experience the "tip of the tongue phenomenon" physically: the name of the thing we wish to recall seems to be literally "out there" like the cars of the hidden half of a ferris wheel, not quite cresting the horizon, that will bring it in view. The brain tells us we have the knowledge somewhere, but the tongue cannot find it. The result in both cases is often a headache, although the child, thankfully, is less likely than we are to worry memory into pain.

Other, more sophisticated views of neurophysiology likewise confirm our intuitive mapping of thought in physical space. Douglas Hofstadter and Daniel Dennett describe such mapping in *The Mind's I:*

> The brain's neurons respond only to "local" stimuli—local in both space and time. At each instant... the neighboring neurons' influences are added together and the neuron in question either fires or doesn't. Yet somehow all of this "local" behavior can add up... to a set of "global" principles that, seen on the level of human behavior, embody long-term goals, ideals, interests, tastes, hopes, fears, morals, and so on. So, somehow all of these long term global qualities have to be coded into the neurons in such a way that, from the neurons' firing, the proper global behavior will emerge. We can call this a "flattening" or "compressing" of the global into the local.[1]

Our thoughts do experience compression and in the process assume a geography in which the global is indeed mapped upon the local, in which ideas literally are part of a physical landscape that we can apprehend proprioceptively, i.e., inwardly and spatially. Consider, e.g., the events of today: waking, the absence of breakfast, conference, travel, talk, weariness, meal, etc. As these events accumulate, merge, and evade articulation, they are unwittingly gathered into something that my brain comes to know and index as "yesterday." Yet we know that most yesterdays fade both from memory and from the physiological landscape with little ado. Thus, it is just as unlikely that I will be able to remember this particular yesterday as it is that I will be able to recall a particular crimson sugar maple among all the autumns of a lifetime. If, however, some more memorable and transcendent index intervenes, the generic yesterday suddenly finds itself not just indexed but also palpably mapped as "the trip," "having an affair," "my father's death," "an exciting time," "the redemptiveness of routine life," or whatever. The Buddha recalls the Bo tree, the lover recalls the crimson sugar maple beneath which he fell into the arms of his first love.

No matter what I attempt to recall tomorrow—whether redemption, my father's death, a maple tree, or a simple yesterday—I have the sense that in recalling it, I am accessing, if not traversing, an actual, and actualized, space. This sense of traversing becomes especially evident if, on some tomorrow, I come to write about what I recall of this particular yesterday. Whatever our critical disputes about the nature of writing, we seem to share a common recognition that, in writing, a mental act becomes a thing, that time, the most primordial of mental acts, yields space, the most primitive of things.

"I take SPACE to be the central fact to man born in America," argues Charles Olson in his meditation on Melville, *Call Me Ishmael*: "I spell it large because it comes large here. Large and without mercy. It is geography at bottom.... That made the first American story... exploration."[2]

Our technological age bears witness to a widespread and growing exploration, in several disciplines, of the possibility of mapping both knowing and recalling, i.e., both the time of experience and the space of memory. In undertaking this exploration of knowing and recalling, it seems useful to test both our efforts and our metaphors against the discipline that most directly considers mapping. It seems useful to seek geographical measures. "A map," says the great U.S. geographer, Carl O. Sauer, "invites attention alike synoptically and analytically...."

Its symbols are translated into images, and these are assembled in the mind's eye into meaningful associations of land and life."[3]

How we map not only our experience but also our recollected knowledge of it informs, and perhaps controls, how we map the actual world of nature. Again, despite violently opposed critical views of the interrelationships among writer, text, and reader, it is possible to fashion a fragile, albeit highly articulated, consensus that the writing process, as we have come to understand it, becomes geographic, and we, like Sauer, seek to describe it in terms of meaningful associations.

That this linkage between mapping mind and mapping nature is not mere metaphor is evident to even a casual reader of current texts in narratology, compositional theory, and any of the varieties of textual criticism, many of which arrive fully outfitted with literal maps of deconstructions, paradigms, syntagma, and so on. In this theoretical landscape the disputes rage not over the existence of the land but also over the nature and agency of the transformation that life works upon it. I.e., we do not dispute the existence of writing but, rather, the transformation that the reader, author, implied author, etc., works upon it.

Jacques Derrida, e.g., would suggest that the central dispute might indeed center upon the question of whether it is possible to separate something called writing from the landscape itself.[4] His essay "Living On" is itself a map, glossed and annotated in a text-long footnote called "Border Lines." In it Derrida argues that the borders of any text disappear into "a differential network, a fabric of traces referring endlessly to something other than itself, to other differential traces": "Thus the text overruns all the limits assigned to it so far."[5]

The evolution of computer writing environments already extends beyond word processors to programs and machines especially suited to mapping writing space and thus capable of tracking the writing experience. E.g., a hypertext "structure editor" such as Storyspace for the Macintosh computer enables and encourages both synoptic and analytic mapping of a writing in graphical form.[6] The writing space visually depicts hierarchical and idiosyncratic organizations of writing elements, and the space itself may be explored and traversed according to expressed descriptions for, and connections between, those writing elements as well as latterly discovered interconnections that emerge within them.

Tools such as Storyspace encourage development of what Sauer calls "the 'morphological eye,' a spontaneous and critical attention to form and pattern [which] every good naturalist has." Because these tools—and, indeed, visually oriented computers like the Macintosh—enable representation of both the writing itself and the form that it assumes, they likewise stimulate attention to morphology, what Sauer calls "the very heart" of a geographer's being:

> We work at the recognition and understanding of elements of form and their relation in function . . . so that we have always to learn about selecting what things are relevant and eliminating the insignificant. Relevance raises the question of why the form is present and how it is related to other forms. Description is rarely adequate and even less often rewarding unless it is tied to explanation. It seems necessary therefore to admit to the geographic bent the fourth dimension of time, interest in how what is studied came to be.[7]

Given the existence of such tools, it is therefore not just possible but necessary to ask in what ways our vision of the actual world is informed by these new ways in which we map our writing. "A language is . . . a horizon," says Roland Barthes in *Writing Degree Zero,* "and style a vertical dimension, which together map out for the writer a Nature, since he does not choose either."[8]

I

Memory, too, is a space we do not choose and that we inhabit as nature. We are by now citizens enough of this closing century to know that questions of the nature of space and the space of nature ultimately reside in the domain of physics. "A space-time map," according to Gary Zukav, in *The Dancing Wu Li Masters,* "shows the positions of things in space and it also shows their positions in time."[9] And yet, while we are able to map space-time in retrospect, "the illusion of events 'developing' in time is due to our particular type of awareness which allows us to see only narrow strips of the space-time picture at a time."[10]

Zukav quotes Louis de Broglie:

> In space-time everything which for us constitutes the past, the present, and the future is given in block. . . . Each observer, as his

> time passes, discovers, so to speak, new slices of space-time which appear to him as successive aspects of the material world, though in reality the ensemble of events constituting space-time exist prior to his knowledge of them.[11]

As writers, we often function despite both our unchosen nature and the preexisting ensemble of space-time events that we witness. We sometimes seem fearless in our ability to process our experiences in writing—more fearless there, in fact, than our lives otherwise allow us to be. We owe this apparent fearlessness, at least in part, to our intuitive sense of the space of our memories. Which is to say that, as writers, we map the world also in block.

Even so, we are likely to feel a shock of recognition if we consider Broglie's "ensemble of space-time events" as a description of the text resulting from a writing process. At end, a text often does seem to be nothing more or less than the actualization of events that exist somewhere prior to our knowledge of them, e.g., Derrida's differential traces. Yet we often want to claim that writing is a process of both discovery and exploration, of creativity.

Thus, by "a commodious vicus of recirculation," we come back to a consideration of the space of memory, what that space could mean within the time of experience that we call writing, and how it might inform our mapping of the natural world. To consider that space is, of course, to construct a map of meaningful associations. The technological arena for this kind of mapping centers on artificial intelligence and cognitive science, in which a number of competing models attempt to map meaningful associations between the time of experience and the space of memory.

Within this arena the current darling of an impatient and entrepreneurial world is the expert system, which is nothing more or less than a map of paths through the known. These paths are bound, and controlled, by a rule base, which purports to choose the "best path" among conflicting choices by acting according to the expert's sense of the placement of memory. Yet this "memory" is domain dependent, and the rules control paths that explore only the discrete memory at hand. Diagnosing symptoms of an illness does not traverse the space of memory occupied by my father's death, although it, too, is owed to illness. Thus, the relevant critique of such a map focuses upon its inability to learn, i.e., to make the kind of meaningful associations (or explanations) that turn illness in my memory not just to my father's death but also to the redemptiveness of routine.

Other models offer potentially richer maps and promise exploration of the unknown paths of learning and creativity. The most interesting among these, from a writer's perspective, emerge from the work of Roger Schank and his colleagues and students at the Yale Artificial Intelligence (AI) Project.[12] Schank is perhaps best known for his conceptual dependency theory and its implementation within "scripts" as a model of episodic memory, although he has since moved well beyond these notions into more complex models of memory organization and learning, including explanations. "Conceptual Dependency," he tells us in *Scripts, Plans, Goals, and Understanding,* "is a theory of the representation of the meaning of sentences," in terms of a small set of "primitive actions," which form a basis for "knowledge structures."[13]

These knowledge structures attempt to model "an episodic view of memory [that] claims that memory is organized around personal experiences or episodes rather than around abstract categories." A script "use[s] specific knowledge to interpret and participate in events we have been through many times" and, therefore, "is a structure that describes appropriate sequences of events in a particular context."[14]

The typical Schankian example is a "restaurant script," which enables us to easily interpret a statement such as "I got a Big Mac on the way to work" without asking the speaker whether she, in fact, paid for this food, left a tip, had wine with her meal, or indeed without asking whether a Big Mac is a person, a disease, and so on. Which is to say that we are able to fill in appropriate details from a sketchy set of particular events. A script is, therefore, less a map of meaningful associations than a context for them, less figure than ground, and Schank immediately moves on to describe larger memory structures that access and use scripts. Even so, this notion of scripted memory provides us with a base line from which to begin mapping meaningful associations with the space of memory and so proposing a naive model of the geography of the word.

Scripted memory may be thought of as the figureless landscape of a beach or meadow. In such a landscape we are likely to maintain only the sketchiest of perceptions in our experience of time, until, i.e., we spy a bright and unusual shell or the meadow gives way to an abyss of hellfire. In these latter cases we are likely to augment our generic mental map of beach or meadow with an utterly specific one of "where I found the conch" or "how I passed into the netherworld." These specifics, which Schank would call remindings (or indices), in fact, actually map memory. More important, they also propose an underlying model of landscape within which a morphological eye may range.

In *Dynamic Memory* Schank considers remindings in the context of a "failure-driven" model of understanding and memory.[15] Memory is failure driven because it gains its features due to our inability to account for them routinely and generically. I remember few meadows routinely, e.g., but likely would always recall the meadow leading to the harrows of hell. Although not literally "always"—for if I remembered all unusual experiences at all times, I would never again be able to take a relaxing walk along the beach or amble through a meadow. Remindings are thus the mental links that at once allow us to forget (or ignore) whole chunks of information in order to get on to new tasks and yet also allow us to recover that same forgotten or ignored information at exactly the time it is appropriate or necessary to do so.

It is important to note here that Schank is nearly alone among artificial intelligence researchers in using the computer as a test bed for expressing representations of thought in order to understand the human mind. Thus, for theorists like Schank, if computer representation of thought is both adequate and accurate, it should be possible to program a computer to utilize that representation to understand, communicate, and recall both the thought itself and its unique features of coherence, variety, individuality, flexibility, and so on. Within models such as these, thought discloses, and thus maps, itself in its connectivity rather than in its substantiveness. Or, to put it differently, all thought begins to disclose itself as narrative, no matter how abstracted, technical, or objective we conceive it.

This view is largely congruent to that proposed by Sauer in "The Morphology of Landscape," in which he defines *landscape* as "a land shape, in which the process of shaping is by no means thought of as simply physical... [but, rather,] a distinct association of forms, both physical and cultural."[16] Landscape for Sauer thus functions much in the way that scripts (and more complex memory organization forms) do for Schank. "The geographer," says Sauer, "may describe the individual landscape as a type or possibly as a variant from type, but he always has in mind the generic and proceeds by comparison."[17]

A geographer's account of a landscape is likewise "not that of an individual scene, but a summation of general characteristics," according to Sauer, and yet "the items selected are based upon 'knowledge of the actual situation,' and there is an attempt at synthesis of the form elements": "Their significance is a matter of personal judgement... [and] the personal element... operates in choosing the qualities to be represented."[18]

Schank's remindings function as mental trails among features made memorable to the degree that they do not fit with the generic model, i.e., the time-space map of the script or other memory organization scheme. So too, for Sauer:

> Geography is distinctly anthropocentric, in the sense of the value or use of the earth to man. We are interested in that part of the areal scene that concerns us as human beings because we are part of it, live with it, are limited by it, and modify it. Thus we select those qualities of landscape in particular that are or may be of use to us.... The physical qualities of landscape are those that have habitat value, present or potential.[19]

This view of landscape, and of the space of memory, is distinctly unlike the traditional Adamic view of dominion over the world. If mappings are narrative in nature—i.e., primarily concerned with the connections among useful entities—then the narrative is one concerned with episodic recognition of successive interdependencies. For this geographer and artificial scientist the world of experience itself announces its usefulness as part of the space of memory, and the dominating Adamic intelligence no longer demands that the world yield to use. Instead, disclosures in time (experience) detail—and, in fact, constitute—memory (space).

An emphasis upon interdependence and opportunism characterizes work proceeding from the Yale AI Project. For instance, the kernel of Natalie Dehn's remarkable model of creativity, a concept she calls "opportunity enhancement meta-goal," seems to argue for the inherent value of what might be called a conceptual innocence (not unlike Keats's notion of "negative capability") upon the part of writers. Dehn's model proceeds from a base line assumption that "forgetting and being reminded of goals is fundamental to intentional behavior" and especially creativity.[20] She characterizes "essential characteristics of creative reasoning" developmentally:

1. sensitivity to unforeseen opportunities, when one has the good fortune for them to arise;
2. willingness to be distracted from what one was doing, if something better comes up;
3. a process of successive reformulation, dramatically increasing the probability that useful opportunities will arise; and
4. having at times a sense of direction that serves both (*a*) to keep the author usefully occupied in progressing toward her goals, and (*b*) to provide new environments in which fortuitous opportunities are likely to arise.

Dehn notes that "this last characteristic, the opportunity enhancement metagoal, accounts for the dual, directed-yet-serendipitous, nature of creative thinking: why conceptual reformulation is so powerful... and how [reminding] can be constructively channelled in the creative thought process encompassing story-generation."[21]

Sauer suggested a similar opportunism as the basis for educating geographers through exploration, i.e., in a "field group as on a journey of discovery, not as a surveying party," noting that, "to some, such see-what-you-can-find fieldwork is irritating and disorderly since one may not know beforehand all that one will find."[22] Even so, he says:

> Locomotion should be slow, the slower the better, and be often interrupted by leisurely halts to sit on vantage points and stop at question marks. Being afoot, sleeping out, sitting about camp in the evening, seeing the land in all its seasons, are proper ways to intensify the experience, of developing impression into larger appreciation and judgement. I know no prescription of method; avoid whatever increases routine and fatigue and decreases alertness.[23]

This may seem romantic, even coming as it does as a valedictory from the father of a discipline, but its intention was tonic, already in opposition (in 1956) to the quantification that "happens to be fostered at the moment by the liking of those who dispense funds for long-term programs and institutional organizations."[24] It is not surprising that thirty years afterward quantification thrives largely for the same reasons. Nor is it surprising that many of Sauer's successors, social scientists and humanists alike, fear that computer models and computer tools only further a regrettable attempt to quantify mind (expert systems seem an instance here). Even so, it is not only possible but necessary to argue that computer tools and computer models can offer a current and effective antidote to a thirty-year (or longer) plague of quantifying. The antidote is, in fact, a worldview, and one not much different from the antidote that Sauer himself suggested in claiming that "we are concerned with processes that are largely non-recurrent and involve time spans mainly beyond the short runs available to enumeration."[25]

Writing is just such a process. Computer tools can help us envision and map both the long span and the nonrecurrent nature of the writing process. More important, in using these tools, we encounter representations of nature that transcend the machine, value interdependence, and focus upon opportunistic, open-minded notions of use.

II

Tools such as Storyspace are to models like Schank's or Dehn's what the naive neurophysiology of a child's hand to the skull is to the complex neurophysiology of the influences of neighboring neurons. Yet, in suggesting an emerging perception of nature, we need not consider something as daunting as the representation of creativity in artificial intelligence or even the less complex ability of a structure editor to map complex connectivities developing over time. We might consider, instead, the increasingly familiar—if not completely commonplace—word processor and the representation of nature that it suggests. To simplify still further, consider a word processor not much more powerful than an intelligent typewriter. Let us say that it allows only the simplest kinds of file management and text manipulation, i.e., addition and deletion of text at any point along the continuum of a text file but without block move, appending, or other simple capabilities.

While such a word processor might not be very useful for any task beyond writing simple notes, it nonetheless makes (and visually presents) certain assumptions about the writing process:

- writers work forward and back in any task;
- writing itself is inwardly elastic, i.e., it grows to accommodate additions and shrinks after deletions;
- writing involves "mistakes" that are physical (typos, misspellings, etc.) and result in minor changes to existing text;
- writing involves mistakes that are structural (intellectual, rhetorical, and other changes) and result in large-scale deletion and substitution in the text.

This example word processor also makes (and less obviously presents) certain assumptions about the structure of writing. These structural assumptions can, more often than not, be attributed to the program's necessary interaction with the programming language, operating system, or hardware design. Even so, they effectively represent structural assumptions:

- that one writing is different from another (most intelligent typewriters do not make this assumption);
- that writings with the same name are not different from one another, even if they are created at different times (although the writer may be able to override this assumption);
- that mistakes (i.e., of either type above, whether single characters or large blocks) need to be either saved or disposed of (depending upon how the program code handles these operations);
- that writing added at a point forward or back in a file is part of the same writing of that file, despite the fact that it is added at different times.

What emerges from these assumptions about process and structure is a strong sense that any writer makes constant choices based upon an initial plan (which may alternately be called a goal, an idea, an intention, or even a story). At the primitive level of this word processor the plan is nothing more than the first word or phrase or, at best, the name of a textfile. A tool for writing is thus seen as a representation of choices. We know, however, that any representation maps itself against a generic plan, i.e., that the map itself is a narrative of divergences from the generic, expressed in terms of episodic connectivities. We can, therefore, ask what difference there is between the narrative of divergence that results from this tool and that which results, say, from an ordinary mechanical typewriter.

The typewriter generally represents choices by retaining the most recent choice, although a writer may, of course, capture other choices by noting alternative words, sentences, or even whole sections, doing so either parenthetically or interlinearly. Thus, the question of how to represent choices might more accurately be said to involve both the extent to which alternatives are represented and the usefulness of the tool in recalling the circumstances of choosing.

Compare the textfile to the typescript using Sauer's terms, i.e., as a "shape, in which the process of shaping is by no means thought of as simply physical... [but rather] a distinct association of forms, both physical and cultural." The textfile in its elasticity readily appears both physically more accommodating and culturally more integrative, and yet it suffers in comparison to the linear typescript to the extent that the textfile is ultimately "seamless," or featureless, with neither physical nor cultural evidences of the details of accommodation and integration that, say, scrawled annotations might present. It is the apparent lack of depth that frightens us in computer tools and makes them seem quantifying.

Yet any capability added to this impoverished word processor dramatically tips the balance in its favor. Block moves of text within the file or date stamping of files within the operating system initially add the potential for immediately capturing depth. Further enhancements such as a windowing capability to view successive drafts or a "dribble file" allowing the writer to undo changes in chronological layers (both of which are now commonplace in current word processors) in a sense allow writers to orchestrate remindings.[26] Structure editors, finally, enable consciously delineated remindings as well as enable search processes to recover developmental ones while simultaneously presenting both the seamless space of the developed writing and the "deconstructed" experience of its creation.

Tools such as these redeem, and rehumanize the dehumanizing proper name, "User," which is computer jargon for *Homo technologicus*. They do so by redefining use in the way Charles Olson did in his concern with the proprioceptive self, i.e., "How to use oneself and on what. . . ." Proprioception is depth perception, from the body (even the body of the text) outward, and nature at this level maps itself against the generic and the quantifying on account of the significance of its details and their usefulness in recalling our experience accurately and appropriately within the intuitive depth of memory. A beach or a meadow has value as the seamless, generic synthesis of form beneath which particular events await their opportunistic expression as an arena for recollecting and using oneself.

The same is true for the textfile that recollects and uses the beach or meadow. Our intuition is that we write proprioceptively, as the child's hand does, summoning the space of memory outward. It is a dream of depth, expressed in the depth of an elastic and windowing world.

"Dreaming depth, we dream our depth," says Gaston Bachelard, the great and dreaming geographer of the word: "Dreaming of the secret power of substances, we dream our secret being. . . . The dream interior is warm, never burning. . . . Through warmth, everything is *deep*. Warmth is the sign of depth, the sense of depth."[27]

"The Ends of Print Culture" (a work in progress)

> Felt is a supple solid product that proceeds altogether differently, as an anti-fabric. It implies no separation of threads, no intertwining, only an entanglement.... [Likewise] patchwork, with its infinite, successive additions of fabric... frees uniquely rhythmic values....
>
> —Deleuze and Guattari 1983a

This essay was originally published in the on-line electronic journal *Postmodern Culture (PMC)* under a much longer name, "Notes toward an Unwritten Non-Linear Electronic Text: 'The Ends of Print Culture' (a work in progress)." It is, as I have noted in the introduction, a true electronic document. This is to say that it uses electronic medium as a space for quilting (or felting) and gathers not just strands but whole sections of nomadic text from earlier writings, entangling them hypertextually into the helix of uniquely rhythmic values. I have neither edited its rather substantial borrowings, other than to italicize them (so a reader who tires of wholesale repetitions can pass them by), nor have I reformatted its text from the numbered paragraphs that serve in lieu of pagination in internet transmission.

[1] *For a period of time last year on each end of our town, like compass points, there was a mausoleum of books. On the north end of town a great remainder warehouse flapped with banners that promised 80 percent off publishers prices. Inside it row upon row of long tables resembled nothing less than those awful makeshift morgues that spring up around disasters. Its tables were piled with the union dead: the mistakes and enthusiasms of editors, the miscalculations of marketing types, the brightly jacketed, orphaned victims of faddish, fickle, or fifteen-minute shifts of opinion and/or history. There an*

appliance was betrayed by another (food processor by microwave); a diet guru was overthrown by a leftist in leotards (Pritikin by Fonda); and every would-be Dickens seemed poised to tumble, if not from literary history, at least from all human memory (already gangs of Owen Meanies leer and lean against faded Handmaidens of Atwood).

[2] *Upon first looking into such a warehouse—forty miles east of our spare parts, Bible Belt midwestern town, in what we outlanders think of as wonderful Ann Arbor, we thought only a university town could sustain this. When the same outfit opened up in our town, and the tables were piled not with the leavings of Ann Arborites but, instead, with towers of the same texts, we knew this was a modern-day circus. Ladies and gentlemen, children of all ages! here come the books!*

[3] *Meanwhile, at the opposite pole in the second mausoleum, a group named the Friends of the Library regularly sell off tables of what shelves can no longer hold. One hundred years of Márquez is too impermanent for the permanent collection of our county library, but so too—at least for the branches that feed pulp back to this trunk—so too is the Human Comedy, so too is the actual Dickens or Emily Dickinson. The book here must literally earn its keep.*

[4] *Both the remainder morgue and the Friends of the Library mortuary are examples of production/distribution gone radically wrong. Books—and films and television programs and software, etc.—have become what cigarettes are in prison, a currency, a token of value, a high-voltage utility humming with options and futures. It is not necessary to have read them. Rather, we are urged to imagine what they could mean to us, or, more accurately, to imagine what we would mean if we were the kind of people who had read them.*

[5] *This is to say that the intellectual capital economy has to some extent abandoned the idea of real, material value for one of utility. This abandonment is not unlike the kind that in a depressed real estate market leaves so-called worthless condos as empty towers in whose shadowy colonnades the homeless camp. Ideas of all sorts have their fifteen-minute Warholian half-life and then dissipate, and yet their structures remain. We have long ago stopped making real buildings in favor of virtual realities and holograms.* The book has lost its privilege. For those who camped in its shadows, for the culturally homeless, this is not necessarily a bad thing. *No less than the sitcom or the Nintendo cartridge, the book too is merely a fleeting, momen-*

tarily marketable, physical instantiation of the network. And the network, unlike the tower, is ours to inhabit.

[6] *In the days before the remote control television channel zapper and modem port we used to think network meant the three wise men with the same middle initial: two with the same last name, NBC and ABC, and their cousin CBS. Now, increasingly, we know that the network is nothing less than what is put before us for use. Here in the network what makes value is, to echo the poet Charles Olson, knowing how to use yourself and on what. Networks build locally immediate value, which we can plug into or not as we like. Thus, the network redeems time for us.* Already with remote control channel zapper in hand, most of us can track multiple narratives, headline loops, and touchdown drives simultaneously across cable transmissions and stratified time. *In the network we know that what is of value is what can be used and that we can shift values everywhere, instantly, individually, as we will.*

[7] *We live in what, in* Writing Space, *Jay Bolter calls the late age of print (1991). Once one begins using a word processor to write fiction it is easy to imagine that the same* technē *that makes it possible to remove the anguish from a minor character on page 251 of a novel manuscript and implant it within a formative meditation of the heroine on page 67 could likewise make it possible to write a novel that changes every time the reader reads it. Yet what we envision as a disk tucked into a book might easily become the opposite. The reader struggles against the electronic book. "But you can't read it in bed," she says—everyone's last-ditch argument. Fully a year after Sony first showed Discman, a portable, mini-CD the size of a Walkman, capable of holding 100,000 pages of text, a discussion on the Gutenberg computer network wanted to move the last ditch a little further. The smell of ink, one writer suggested; the crinkle of pages, suggested another.*

[8] *Meanwhile, in far-off laboratories of the Military-Infotainment Complex—to advance upon Stuart Moulthrop's phrase (1989b)—at Warner, Disney, or IB-Apple and MicroLotus, some scientists work on synchronous smell-o-vision with real-time simulated fragrance degradation shifting from fresh ink to old mold, while others build raised-text touch screens with laterally facing windows that look and turn like pages, crinkling and sighing as they turn. "But the dog can't eat*

it," someone protests, and, smiling silently, the scientists go back to their laboratories, bags of silicon kibbles over their shoulders.

[9] *What we whiff is not the smell of ink but, rather, the smell of loss: of burning towers or men's cigars in the drawing room. Hurry up, please—it's time. We are in the late age of print; the time of the book has passed. The book is an obscure pleasure like the opera or cigarettes. The book is dead—long live the book. A revolution enacts what a population already expresses: like eels to the Sargasso, a hundred thousand videotapes annually return to a television show about home videos. In the land of polar mausolea, in this late age of print, swimming amid this undertow, who will keep the book alive?*

[10] In an age when more people buy and do not read more books than have ever been published before, often with higher advances than ever before, perhaps we will each become like the living books of Truffaut's version of Bradbury's *Fahrenheit 451,* whose vestal readers walk along the meandering river of light just beyond the city of text. We face their tasks now, resisting what flattens us, reembodying reading as movement, as an action rather than a thing, network out of book.

[11] We can reembody reading if we see that the network is ours to inhabit. There are no technologies without humanities; tools are human structures and modalities. Artificial intelligence is a metaphor for the psyche, a contraption of cognitive psychology and philosophy; multimedia (even as virtual reality) is a metaphor for the sensorium, a perceptual gadget beholding to poetics and film studies. Nothing is quicker than the light of the word. In "Quickness," one of his *Six Memos for the Next Millennium,* Italo Calvino writes:

> In an age when other fantastically speedy, widespread media are triumphing and running the risk of flattening all communication onto a single, homogeneous surface, the function of literature is communication between things that are different simply because they are different, not blunting but even sharpening the differences between them, following the true bent of the written language. (1988, 45)

[12] Following the true bent of the written language in the late age of print brings us to the topographic. "The computer," Jay Bolter says, "changes the nature of writing simply by giving visual expres-

sion to our acts of conceiving and manipulating topics." In the topographic city of text, shape itself signifies, as in Warren Beatty's literally brilliant rendering of the city of Dick Tracy. There the calm, commercial runes of marquee, placard, neon, and shingle (DRUGS, LUNCHEONETTE, CINEMA) not only map the pathways of meaning and human intercourse but also shape and color the city itself and its inhabitants. Face and costume, facade and meander, river's edge and central square, booth or counter, Trueheart or Breathless. "Electronic writing," says Bolter, "is both a visual and verbal description":

> It is not the writing of a place, but rather a writing with places, spatially realized topics. Topographic writing challenges the idea that writing should be merely the servant of spoken language. The writer and reader can create and examine signs and structures on the computer screen that have no easy equivalent in speech. (1991, 25)

[13] *Ted Nelson, who coined the term* hypertext *in the 1960s, more recently defined it as "non-sequential writing with reader controlled links." Yet this characterization stops short of describing the resistance of this new object. For it is not merely that the reader can choose the order of what she reads but that her choices in fact become what it is.*

[14] *Let us say, instead, that hypertext is reading and writing electronically in an order you choose—whether among choices represented for you by the writer or by your discovery of the topographic (sensual) organization of the text. Your choices, not the author's representations or the initial topography, constitute the current state of the text.* You become the reader-as-writer.

[15] We might note here that the word we want to describe the reader-as-writer already exists, although it is too latinate and bulky for contemporary use. *Interlocutor* has the correct sense of one conversant with the polylogue as well as the right degrees of burlesque, badinage, and bricolage behind it. Even so, we will have to make do with, and may well benefit by extending, the comfortable term *reader.*

[16] We may distinguish two kinds of hypertext according to their actions (Joyce 1988). Exploratory hypertext, which most often occurs in read-only form, allows readers to control the transformation of a

defined body of material. It is perhaps the type most familiar to you, if you have seen a Hypercard stack. (Note here that a stack is the name of the electronic texts created by this Apple product. There are other hypertext systems, such as Storyspace and Supercard for the Macintosh, or Guide for both the Macintosh and MS-DOS machines, and the newcomer ToolBook for the latter.)

[17] In the typical stack the reader encounters a text (which may include sound and graphics, including video, animations, and what have you). She may choose what and how she sees or reads, either following an order the author has set out for her or creating her own. Very often she can retain a record of her choices in order to replay them later. More and more frequently in these documents she can compose her own notes and connect them to what she encounters, even copying parts from the hypertext itself.

[18] This kind of reading of an exploratory hypertext is what we might call empowered interaction. The transitional electronic text makes an uneasy marriage with its reader. It says: you may do these things, including some I have not anticipated.

[19] It is to an extent true that neither the author's representations nor the initial topography but, instead, the reader's choices constitute the current state of the text for her. In these exploratory hypertexts, however, the text does not transform or rearrange itself to embody this current state. The transitional electronic text is as yet a marriage without issue. Each of the reader's additions lies outside the flow of the text, like Junior's shack at the edge of the poster-colored city of Dick Tracy. The text may be seen as leading to what she adds to it, yet her addition is marginal, ghettoized. *Stuart Moulthrop suggests that to the extent that hypertexts let a power structure "subject itself to trivial critiques in order to pre-empt any real questioning of authority, . . . hypertext could end up betraying the anti-hierarchical ideals implicit in its foundation" (1989a, 21). Under such circumstances the reader's interaction does not reorder the text but, rather, conserves authority. She moves outside the pathways of meaning and human intercourse, unable to shape and color the city itself or its inhabitants.*

[20] Even so, to the extent that the topographical writing of an exploratory hypertext lets readers create and examine signs and structures, it does make implicit the boundary that both marks and makes

privilege or authority. In fact, it has always been true that the interlocutory reader, let us say brooding alone in the reading room of the British Museum, might come to see this boundary. Attuned to organizational structures of production and reproduction, she might mark with Althusser, "the material existence of an ideological apparatus" of the state (1971).

[21] But she might not be able to see quite as clearly or as quickly as she can see in the hypertext how the arena is organized to marginalize and diminish her. This is the trouble with hypertext, at any level: it is messy; it lets you see ghosts; it is always haunted by the possibility of other voices, other topographies, others' governance.

[22] Print culture is as discretely defined and transparently maintained as the grounds of Disney World. There is no danger that new paths will be trod into the manicured lawns. Some would like to think this groundskeeping is a neutral decision, unladen, decontextualized, removed from issues of empowerment, outside any reciprocal relationship. *For the moment institutions of media, publishing, scholarship, and instruction depend upon the inertia of the aging technology of print, not just to withstand attack on established ideas but also to withstand the necessity to refresh and reestablish these ideas. In fact, hypermedia educators frequently advertise their stacks by featuring the fact that the primary materials are not altered by the webs of comments and connections made by students. This makes it easier to administer networks, they say.*

[23] Like the Irish king Cuchulain who fought the tide with his sword, they lose who would battle waves on the shores of light. The book is slow, the network is quick; the book is many of one, the network is many ones multiplied; the book is dialogic, the network polylogic.

[24] The second kind of hypertext, constructive hypertext, offers an electronic alternative to the gray ghetto alongside the river of light. *Constructive hypertext requires a capability to create, change, and recover particular encounters within a developing body of knowledge. Like the network, conference, classroom, or any other form of the electronic text, constructive hypertexts are "versions of what they are becoming, a structure for what does not yet exist" (Joyce 1988).*

[25] As a true electronic text, the constructive hypertext differs from the transitional exploratory hypertext in that its interaction is

reciprocal rather than empowered. The reader gives birth to the true electronic text. It says: what you do transforms what I have done and allows you to do what you have not anticipated. "It is not just that [we] must make knowledge [our] own," says Jerome Bruner in *Actual Minds, Possible Worlds,* but that we must do so "in a community of those who share [our] sense of culture" (1986, 127).

[26] *A truly constructive hypertext will present the reader with opportunities to recognize and deploy the existing linking structure in all its logic and nuance. I.e., the evolving rhetoric must be manifest for the reader. She should be able to extend the existing structure and to transform it, harnessing it to her own uses. She should be able to predict that her own transformations of a hypertext will cause its existing elements to conform to her additions. While not merely taking on but surrendering the forefront to the newly focused tenor and substance of the interlocutory reader, the transformed text should continue to perform reliably in much the same way in which it has for previous readers.*

[27] Indeed, every reading of the transformed text should in some sense rehearse the transformation made by the interlocutory reader. If a reader, let us call her Ann, has read a particular text both before and after the intervention of the interlocutory reader, Beatrice, Ann's experience of the text should have the familiar discomfort of recognition. Ann should realize Beatrice's reading.

[28] Not surprisingly, the first efforts at developing truly constructive hypertexts have taken place in (hyper)fictions. *afternoon* (Joyce 1990) attempts to subvert the topography of the text by making every word seem as if it yields other possibilities, letting the reader imagine her own confirmations. This "letting" likely signifies a partially failed attempt, a text that empowers more than it reciprocates. In situating and criticizing *afternoon,* Stuart Moulthrop imagined "a writing space [that] presumes a new community of readers, writers, and designers of media... [whose] roles would be much less sharply differentiated than they are now" (1989a).

[29] In attempting to develop such a community, it becomes clear to hyperfiction writers that, unless roles of author and reader are much less sharply differentiated, the silence will have no voice. Even interactive texts will live a lie. "In all claims to the story," writes the Canadian poet Erin Mouré,

> ... there is muteness. The writer as witness, speaking the stories, is a lie, a liberal bourgeois lie. Because the speech is the writer's speech, and each word of the writer robs the witnessed of their own voice, muting them.
>
> —1989, 84

[30] Increasingly, hyperfiction writers consider how the topographic (sensual) organization of the text might present reciprocal choices that constitute and transform the current state of the text. How, in the landscape of the city of text, can the reader know that what she builds will move the course of the river? How might what she builds present what Bruner calls an invitation to reflection and culture creating? In her poem "Site Glossary: Loony Tune Music" Mouré says that,

> witness as a concept is outdated in the countries of
> privilege, witness as tactic, the image as completed
> desktop publishing & the writer as accurate, the names are
> sonorous & bear repeating tho there is no repetition the
> throat fails to mark the trace of the individual voice which
> entails loony tune music in this age.
>
> —1989, 84

[31] Hyperfictions seek to mark the trace with their own loony tune music. In *Chaos* Stuart Moulthrop has speculated a fiction that is consciously unfinished, fragmentary, open, one of emotional orientations and transformative encounters. John McDaid's hyperfiction *Uncle Buddy's Phantom Fun House* is an electronic world of notebooks, scrap papers, dealt but unplayed Tarot cards, souvenirs, segments, drafts, and tapes, unfinished in the way that death unfinishes us all (1991). In *Izme Pass,* their hyperfictional "deconstruction of priority," Carolyn Guyer and Martha Petry seek "to weave ... [a] new work made not of the parts but the connections ... [in order] to unmurk it a little, to form connection in time and space, but without respect to those constraints" (Guyer and Petry1991b).

[32] While this may seem the same urge toward a novel that changes each time it is read, what has changed in the interim between novelist-at-word-processor and hyperfiction writer is that computer tools to accomplish these sorts of multiple texts have been built. Moreover, hyperfiction writers have not only imagined and rendered them

but also, and more important, have begun to set out an aesthetic for a multiple fiction that yields to its readers in a reciprocal relationship.

[33] This sort of reciprocal relationship for electronic art has a conscious history in the late twentieth century. In "Strauss and the Electronic Future" Glenn Gould envisions a "multiple authorship responsibility in which the specific functions of the composer, the performer, and indeed the consumer overlap" (1964, 33). He expands this notion in his extraordinary essay "The Prospects of Recording": "Because so many different levels of participation will, in fact, be merged in the final result, the individualized information concepts which define the nature of identity and authorship will become very much less imposing" (1966, 14).

[34] *What joins the concerns of many of the writers working with multiple fictions is nothing less than the deconstruction of priority involved in making identity and authorship much less imposing. "The fact in the human universe," says Charles Olson, "is the discharge of the many (the multiple) by the one (yrself done right... is the thing—all hierarchies, like dualities, are dead ducks)" (1974, 39).*

[35] *These writers share a conviction that the nature of mind must not be fixed. It is not a transmission but, rather, a conversation we must keep open. "If structure is identified with the mechanisms of the mind," says Umberto Eco, "then historical knowledge is no longer possible" (1989). We redeem history when we put structure under question in the ways that narrative, hypertext, and teaching each do in their essence. Narrative is the series of individual questions that marginalize accepted order and thus enact history. Hypertext links are no less than the trace of such questions, a conversation with structure. All three are authentically concerned with consciousness rather than information, with creating and preserving knowledge rather than with the mere ordering of the known. The value produced by the readers of hypertexts or by the students we learn with is constrained by systems that refuse them the centrality of their authorship. What is at risk is both mind and history.*

[36] In Wim Wenders's (and Peter Handke's) film *Wings of Desire* the angels walk among the stacks and tables of a library, listening to the music within the minds of the individual readers. It is a scene of indescribable delicacy and melancholy both (one that makes you want to rush from the theater and into the nearest library, there to read

forever), into the midst of which, shuffling slowly up the carpeted stair treads, huffing at each stairwell landing, his nearly transparent hand touching on occasion against the place where his breastbone pounds beneath his suit and vest, comes an old man, his mind opening to an angel's vision and to us in a winded, scratchy wheeze.

[37] "Tell me muse of the storyteller," he thinks, "who was thrust to the end of the world, childlike ancient. . . ." The credits tell us later that this is Homer. "With time," he thinks, "my listeners became my readers. They no longer sit in a circle; instead, they sit apart and no one knows anything about the other. . . ."

[38] Homer's is for us, increasingly, an old story. *When print removed knowledge from temporality, Walter Ong reminds us, it interiorized the idea of discrete authorship and hierarchy. Ong envisioned a new orality (1982).* In this case it is a film that restores the circle; likewise, the "multiple authorship" of hypertext offers an electronic restoration of the circle.

[39] Although *hypertext* is an increasingly familiar cultural term, its artistic import is only beginning to be realized—in novels whose words and structures do not stay the same from one reading to another, ones in which the reader no longer sits apart but, by her interaction, shapes and transforms.

[40] *Shaping ourselves, we ourselves are shaped. This is the reciprocal relationship. It is likewise the elemental insight of the fractal geometry: that each contour is itself an expression of itself in finer grain. We have been talking so long about a new age, a technological age, an information age, that we are apt to forget that it is we who fashion it, we who discover and recover it, we who shape it, we who literally give it form with how we use ourselves and on what.*

[41] This organic reconstitution of the text may be what makes constructive hypertext the first instance of what we will come to conceive as the natural form of multimodal, multisensual writing: the multiple fiction, the true electronic text, not the transitional electronic analogue of a printed text like a hypertextual encyclopedia. Fictions like *afternoon, WOE, Victory Garden, IZME PASS, Quibbling,* or *Uncle Buddy's Phantom Funhouse* can neither be conceived nor experienced in any other way. They are imagined and composed within their own idiom and electronic environment, not cobbled together from preordained texts.

[42] For these fictions there will be no print equivalent nor even a mathematical possibility of printing their variations. Yet this is in no way to suggest that these fictions are random, on the one hand, or display artificial intelligence, on the other—merely that they are formational.

[43] What they form are instances of the new writing of the late age of print, what Jane Yellowlees Douglas terms "the genuine post-modern text rejecting the objective paradigm of reality as the great 'either/or' and embracing, instead, the 'and/and/and'" (1991, 125). The issues at hand are not technological but, rather, aesthetic, not what and where we shall read but how and why. These are issues that have been a matter of the deepest artistic inquiry for some time and share a wide and eclectic band of progenitors and a century or more of self-similar texts in a number of media.

[44] The layering of meaning and the simultaneity of multiple visions have gradually become comfortable notions to us, though they form the essence underlying the intermingled and implicating voices of Bach, which Glenn Gould heard with such clarity. We are the children of the aleatory convergence. Our longing for multiplicity and simultaneity seems upon reflection an ancient one, the sole center of the whirlwind, the one silence.

[45] It is an embodied silence that the multiple fiction can render. We find ourselves at the confluence of twentieth-century narrative arts and cognitive science as they approach an age of machine-based art, virtual realities, and what Don Byrd calls "proprioceptive coherence" (1991). The new writing requires rather than encourages multiple readings. It not only enacts these readings; it does not exist without them. Multiple fictions accomplish what its progenitors—lacking a topographic medium, light speed, electronic grace, and the willing intervention of the reader—could only aspire to.

INTERSTITIAL

Artists' Statements— Giving Way(s) before the Touch

afternoon, a story is created with Storyspace, a hypertext program that is both an author's tool and a reader's medium.

- You move through the text by pressing the Return key to go from one section to another (i.e., "turn pages"), and you click the Back arrow (on the toolbar) to go back ("page back");
- You double-click on certain words to follow other lines of the story. Window titles often confirm words that "yield."

The story exists at several levels and changes according to decisions you make. A text you have seen previously may be followed by something new, according to a choice you make or already have made during any given reading.

✪

I haven't indicated what words yield, but they are usually ones that have texture as well as character names and pronouns.

There are more such words early on in the story, but there are almost always options in any sequence of texts. The lack of clear signals isn't an attempt to vex you, rather an invitation to read either inquisitively or playfully and also at depth.

✪

In my mind the story, as it has formed, takes on margins. Each margin will yield to the impatient, or wary, reader. You can answer yes at the beginning and page through on a wave of Returns or page through directly, again using Returns, without that first interaction.

These are not versions but the story itself in long lines. Otherwise, however, the center is all—Thoreau or Brer Rabbit, each preferred the bramble. I've discovered more there too, and the real interaction, if that is possible, is in pursuit of texture.

There we match minds.

✪

Closure is, as in any fiction, a suspect quality, although here it is made manifest. When the story no longer progresses or when it cycles or when you tire of the paths, the experience of reading it ends. Even so, there are likely to be more opportunities than you think there are at first. A word that doesn't yield the first time you read a section may take you elsewhere if you choose it when you encounter the section again, and sometimes what seems a loop, like memory, heads off again in another direction.

There is no simple way to say this. **(1987)**

I cannot say that I have not been waiting to be asked what prompted *afternoon, a story* or what it seems to me to be as an art object, an electronic construction, a constant occasion, Holzer's chasing light high above the town square of the city of text, nomadic shiftings of momentary truths like sand flitting from the stone shelf in the wind, a story that changes each time you read it—"a form for what does not yet exist." Thus, I am happy to point to what is perhaps too obvious to be seen: that the screen gives (way) before & with & after the touch, its surface not so much mottled as smoothed in the "places that yield" (the "words with texture," as the reading "directions" say, themselves nothing less than exhibit notes). What kind of text has directions? the hypnotic, the Eventualists' stimulus framed in the mirrored eye (viz. Lombardo 1979), the shape of the mind later seen in dreams, the text of water where Bridal Veil Falls (Basho's Urami-no-Taki) smooths the sandstone shelf in one hundred million drumming fingers of light.

Print stays itself; electronic text replaces itself. If with the book we are always printing—always opening another text unreasonably composed of the same gestures—with electronic text we are always painting, each screen unreasonably washing away what was and replacing it with itself. The eye never rests upon it, though we are apt to feel the finger can touch it. The feel for electronic text is constant and plastic, the transubstantiated smear that, like Silly Putty, gives way to liquid or, like a painter's acrylics, forms into still encapsulated light. We are always painting. The electronic is not at all the touch of the uncertain reader, who—like a child poking at a line of ants or lining up raisins—runs a finger along each cast line of print. Rather, it is certain touch, like holding moths, feeling the velvety resistance as the wing's scales slip off against the press of the fingers, dusting the whorls of the fingertip at whose root the hooked tangle of a zipper row of pincers clings, below that, next to the clasped, now transparent wings, dangles the segmented leathery body, the husk of the idea dangling as palpable as Plato's dry forms, yet a worm still wet within, exploding with damp, phosphor light should you squish it open.

I wanted to create text that gave way(s) before the touch, that could be caressed into motion or repose without end. It began with a brown composition: a compact lunch (cylinder of apple juice, chevron stack of two corned beef halves on rye, tube of Dijon, rectangle of Heath bar) brought to me by a slim-hipped, taupe woman I know only as a mystery of texture. Thus, the initial mystery was a problem in formal composition: "You think of me as brown not ice" the character says, texture becoming your text becoming textual yearning. (The Italian art critic Miriam Mirolla dreamed and drew herself with chocolate lipstick after the SPECCHIO TACHISTOSCOPICO and "was happy for the sweetness of it.") *afternoon* is likewise what follows from what we see in ourselves, the morning after a dark dream: "I want to say that I may have seen my son die this morning," the second screen begins.

Hypertext Narrative

Hypertext is, of course, young by almost any measure—at the time of this writing (1992) 47 years since Vannevar Bush's Memex, 29 since Engelbart's Augment, 27 since Nelson's hypertext, and 5 years since a number of us gathered for the first of the ACM hypertext meetings in 1987.

Narrative is old. Historically, in the Sumer narrative poems joined warehouse inventories as the first texts known. This is quite fitting since, as Ezra Pound has argued, "the artist is one of the few producers": "He, the farmer, and the artisan create wealth; the rest shift and consume it" (1954).

Narrative creates wealth. Yet hypertext, thus far at least, mimics most computational systems in shifting and consuming rather than creating wealth. This is because hypertext thinks itself to be structural rather than serial thought—thought in space rather than thought for space. "Serial thought," says Umberto Eco, "aims at the production of history and not at the rediscovery—beneath history—of the atemporal abscissae of all possible communication" (1989, 221).

Hypertext, at least when it is seen as constructive rather than exploratory, is serial thought. Its "mode of spatialization," Deleuze and Guattari's term, is being *for* space, what I call the constructive, a form for what does not yet exist, rather than being *in* space, or the exploratory and colonizing. Like AI, its elder sister in this science, hypertext would like to claim its creative powers but to do so must become the art it claims it is.

"Look," we want to say of the hypertext, "see this traversal, this web, this trail, this graph.... No one has ever thought this before. Look how, like pearls on a string, these nodes glitter on their path...."

But always some emperor-baiting child within us will out. "Pardon," she says, "But why, if this is new, is it there already? Surely you mean no one has gotten to this point in this way before; surely that is what is new."

"No, child," we say, "You don't understand. This substance is an object of value unto itself. . . ."

Then she asks the terrifying questions: "Who is the author of this new thing? What will you call it? And where will you put it?"

For the reader of a *technical* communication, theoretically at least, there *is* some nethermost node, a gleaming target that represents the meaning of a text. (In reality, of course, this node likely encompasses the span of a series of nodes.) This theoretical terminal node can be systematically described by both its location and its links. Even if the meaning is potentiated, as an inference or discovery, e.g., the node is thought to be present and reachable. I.e., every reading by every reader is thought to be anticipated by the system of the exploratory text. We might call this belief the myth of emerging order (fig. 1).

Within the mythic system of emerging order the reader's task is to make meaning by perceiving order in space. This characteristic Eco identifies with structural thought: "The aim of structural thought is to discover, whereas that of serial thought is to produce" (1989, 221). In the exploratory hypertext interaction is recognition; the system of the text is the author of its additions.

Should the reader wish to add her perceived meaning to the document, there is theoretically no impediment, only a question of privi-

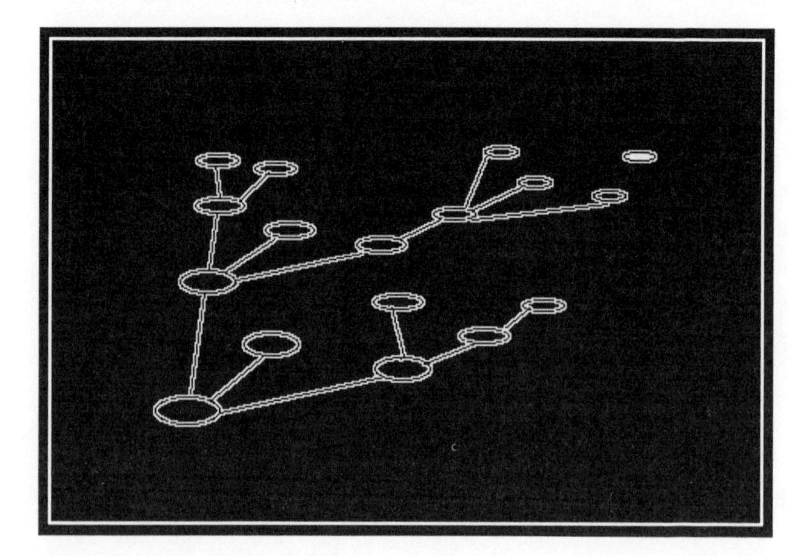

Fig. 1. ". . . The addition is perpetually marginal"

lege and location. We determine the reader's privilege and judge the utility of the addition according to the match we perceive between the reader's order and what we take to be inherent order. We ask *if* a reader can add to the document: Is the summary unique? the inference supported? the discovery consonant? I.e., is this addition authorized?

We likewise determine *where* a reader can add to the document. Once you are privileged you can write into a technical document if its structure accounts for what you add. The document is seen as the atomic shell within which the valency of each addition finds its weighted place. The addition is a terminal node; the text may be seen as leading to it. Yet the addition is perpetually marginal. Interaction does not reorder the text but, rather, conserves authority. Reordering requires a new text, new authority.

The myth of emerging order seems especially instrumental in maintaining technical hypertexts. Because technical hypertexts are thus far treated like centralized rather than distributed knowledge (manuscripts rather than books), individuated meaning must be marginalized and inherent order privileged. Authority, because it is given, is ungained. Interaction remains a utility for the individual reader, i. e., annotation. Hypertext programming—in the double sense of instructed machine behavior and information content: the thing that, like situation comedy, weather map, or docudrama, is shown on the screen—is privileged, centralized, and self-sufficient.

Hypertext narrative transposes this mythic system. If a technical reading is marginal to its text, the text of a narrative is marginal to its reading. Technical communication is governed by the myth of emergent order, i.e., a belief that the structure of meaning emerges from the text. Hypertext narrative produces the present-tense contour of meaningful structures. Meaning in narrative is an orderly but continual replacement of meaningful structures throughout the text. "Although conventional reading habits apply within each lexia," says George P. Landow—borrowing this term for the "nodes and links" from Barthes—"once one leaves the shadowy bounds of any text unit, new rules and new experiences apply" (1992, 4). We judge an addition to narrative by asking if the story continues to "read right," a reader's question, instead of how in technical communication one asks if it fits.

Self-sufficient meaning is not generally characteristic of nontrivial narratives. Only riddles or games, and few of those, offer a target node that truly holds their meaning. In most cases meaning literally has no place in narrative; the meaning of narrative is not in its space but, rather, exists for the space of its unfolding. The moral of a story that has one is part of its story, not its meaning. Meaning in narrative is always strictly potentiated—outside and in some sense prior to the text. It is unreachable and thus not yet linked.

"The fable unwinds from sentence to sentence, and where is it leading?" writes Italo Calvino, "To the point at which something not yet said, something as yet only darkly felt by presentiment, suddenly appears and seizes us and tears us to pieces" (1982, 18). Every reading by every reader becomes privileged and authorized. Reading becomes (and therefore alters) the system of the text.

The reader cannot avoid adding her own meaning to the document. Addition is not merely privileged but is, in fact, required in order to span to the nethermost target node outside the text. Interaction is perception of the principles of ordering. Because hypertexts are read where they are written and written as they are read, interaction is the assumption of authority over the replacement of one writing by another. The reader's orderly perception leads to the placement of an apparent meaning.

Under such circumstances each addition is an initial node to the replaced story she tells in her reading. The target node becomes, in-

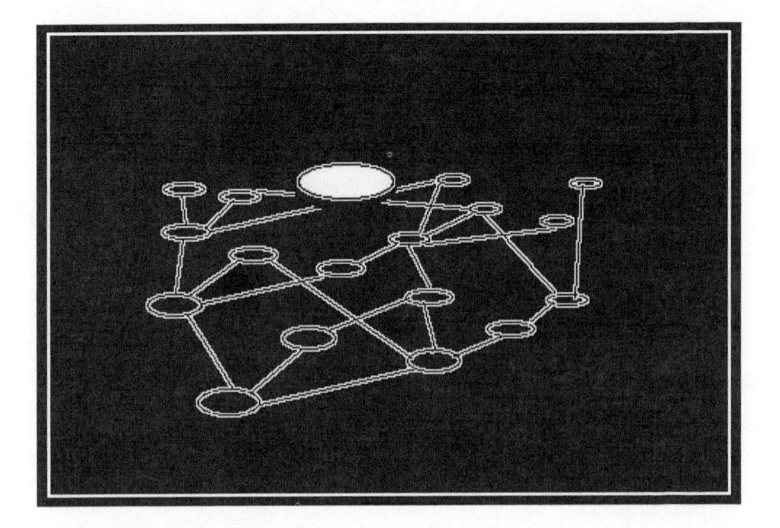

Fig. 2. "... The text itself becomes marginal; meaning reorders the text"

stead, a source node; the text may be seen as leading from it. To accommodate the location of the addition, the text itself becomes marginal; meaning reorders the text. The meaning of a story that had none becomes part of its story. Reordering requires a new text; every reading thus becomes a new text.

Despite these heroic claims, readers are thus far not particularly apt to participate in maintaining, i.e., coauthoring, narrative hypertexts. Coauthorship, rather than the Disneyite laurels of perceived first-personhood, is the true index of interaction. It is not the case that a reader can only interact if she is convinced she is in her own idea space or story world. Millennia of human experiences of empathy—not to mention the most recent one hundred years of psychobiology—not only suggest that I can be lost in your story or your idea but also that I am lost if I do not make it my own. It is in this sense that the four principles of the TINAC Dryden Statement (1988) suggested a future text:

1. **No interruptions:** Reading should be a seamless and uninterrupted experience. Its choices proceed from the expression of possibilities as a narrative medium and depend upon the complicity of the reader in the creation of a narrative. Reading is design enacted.

2. **Any person:** Interaction manifests itself through recognition, sympathy, and witness as much as through impersonation, perception, and exploration. Apprehension of character is participatory design.

3. **Every ending:** Closure, as it has been described by Andrew Sussman, is "the completion of self by the reader." It is, in this sense, design determined.

4. **A read-write revolution:** Interactive narratives are "evolving narratives," written, whether by reader or writer. Authorship is an invitation to active design.

Hypertext narratives become virtual storytellers, and narrative is no longer disseminated irreversibly from singer to listener or writer to reader. It exists, instead, as a cycle in which readers become coauthors and artificially intelligent systems "read" their responses. At the moment most interest centers on the machine side of this cycle, focusing on hypertext or virtual reality systems and intelligent agent/navigators who can structure or generate narratives. Even in a system that generates narrative, however, interaction is the assumption of authority over the replacement of one writing by another. Thus, the human side of the cycle demands primary attention.

To do so, the primacy of the text must be marginalized. The reader declares independence from the software agent and the contingent structure of the virtual reality alike. If the reader is programming, the reader is programmed. Hypertext narrative asserts the authority of the individual reader, no longer privileged and centralized by the system of the text but, rather, by her reading itself. "Once we have dismantled and reassembled the process of literary composition, the decisive moment will be that of reading" says Calvino, "even though entrusted to machines, literature will continue to be a 'place' of privilege within human consciousness, a way of exercising the potentialities contained in the system of signs belonging to all societies at all times" (1982, 16).

There is a danger in asking too much of a technology in its early years—consider AI—but there is a similar danger in allowing a technology to mature unchecked. Change is incremental but also instrumental: it defines as it refines. Hypertext is young; narrative is old. In its preadolescence hypertext attempts the impossible: a preservation of hierarchy through managed individuality. It is consumerist at base. Desire given the form of floor plan, the shopping mall, or living room as loci of being in space.

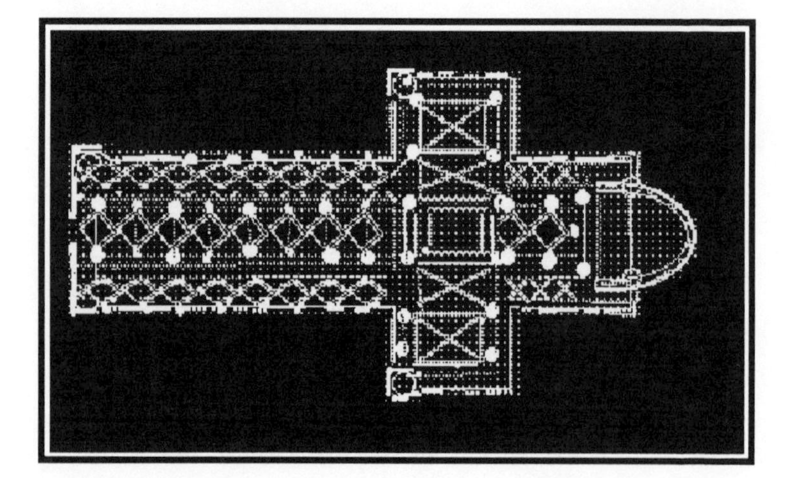

Fig. 3. "... Something like a shopping mall"

In the marketplace, at least, hypertext for now remains unsurprisingly a process of shifting and consuming value within a known map and body of material. Whether cloaked as the "Multimedia Solution" or as part of the jingly joy of "everybody's doin' it, PS2ing it!" our age is one in which, as Hakim Bey characterizes it, "speed and 'commodity fetishism' have created a tyrannical false unity which tends to blur all cultural diversity and individuality so that 'one place is as good as another'" (1991, 107). Our image of hypertext as multimedia is something like a shopping mall, itself the Platonic form of that form of a graph known as a hierarchical tree. The reader is a consumer on a shopping spree, a squirrel leaping branches, while on a high branch Peter Jennings plays the part of that wise nocturnal bird, Scully's Navigator embodied as the software agent called the Reporter (*MacWeek* 6, no. 9 [1992]).

Our image of hypertext as virtual reality is the wire-frame living room of cyberspace decorated with overstuffed tulip-back wire-frame chairs (fig. 4). Its ripstop, crayola-colored upholstery is dithered and fractal nubbed. Squatting in air, one is seated in the stretched, simulated man-made fiber; thereafter, it is possible to copulate with a lobster, examine a cancer cell, joust with a medieval man, or fly up to the ceiling fixture by moving an imaginary stick. The reader is furniture mover in the carefully modeled three-dimensional space of a fractal furniture warehouse, swatting at mosquito vectors in a grid of 3D sound.

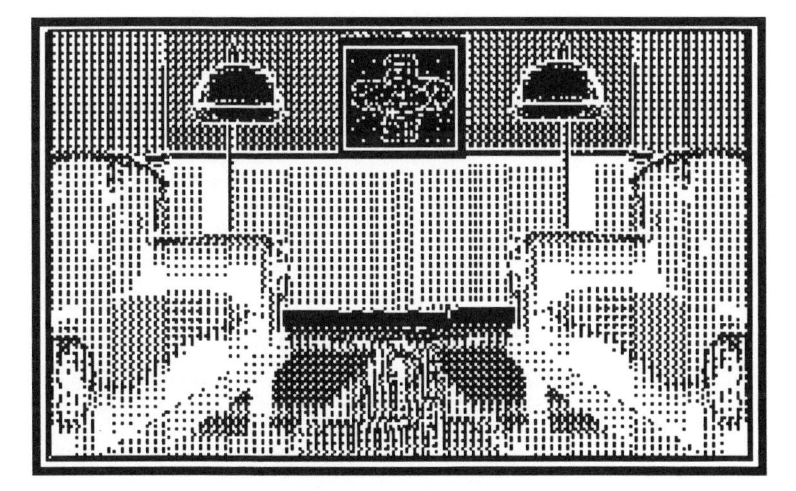

Fig. 4. "... The wire-frame living room of cyberspace"

The warehouse is the familiar locale of most avant-garde theater, although this time the props and flats are rendered in light. A virtual universe of attention has been given to modeling what in old-fashioned narrative one called setting and mood. Tediously earnest rule bases attempt to log precepts of Aristotle, Brecht, and Propp in order to generate lovable Fujisney software agent action figures, which in old-fashioned narrative one called characters. Yet not enough attention has been given to understanding narrative at the interstices, i.e., beneath the virtual seat cushion where a jumble of spilled tokens, sesame seeds, abracadabra apparatus, and salt changes lobsters to princes, transfigures light fixtures to nebulas, and makes medieval jousters weep in sun-bleached parlors.

Jay Bolter suggests that AI is nothing more or less than a machine instantiation of drama. "Artificial intelligence programmers are themselves engaged in a perpetual search for the author," he writes: "They will admit to being the authors of their programs but they refuse to take credit for the output.... If the program writes a story... they look for the author in the program itself... which for them constitutes intelligence." Yet, as such, AI is potentially an art that women and men and their machines (but not machines themselves) can use to create value. "At all times the computer, like all previous technologies of writing," says Bolter, "is intelligent only in collaboration with human readers and writers" (1991, 191, 193).

We make a like claim for hypertext narrative, whether as Eastgate reader, expanded book, multimedia, or virtual reality. In doing so, however, we must own up to our responsibility—even pledge our fidelity—to the forlorn, misunderstood, and lonesome machines who increasingly live both for and with us. We are so dazzled by the arrival of the cyborg at the eternal costume ball that we begin to think it can dance barefoot and without us. We form ourselves in instamatic memory and camcorder consciousness, situate ourselves upon the hypertextual subconscious beneath the interstitial seat cushions. Snapshots don't watch themselves, nor do stories tell themselves. Narrative is not an application but, rather, the ur-system of hypertext. Our emperor-baiting child asks: "Who is the author of this new thing? What will you call it? And where will you put it?" Doing so, she uncovers her disguise, disclosing herself in her three aspects as our old friend and goddess, *Mousa ex machina*. The muse child is revealed in her aspect as the old woman; her question is one of consciousness; it is asked and answered in the asking by mother narrative herself: You are the author, she murmurs; you call the new work history, and you place it in your mind, your mind in it.

What Happens as We Go?

Hypertext Contour, Interactive Cinema, Virtual Reality, and the Interstitial Arts of Jeffrey Shaw and Grahame Weinbren

THE LEGIBLE CITY . . . [is] a three dimensional database [which] represents a virtual space approximately 2.5 kilometres square where the viewer can travel using a bicycle. In the first version, the ground plan was based on an area of Manhattan south of Central Park. Instead of buildings lining the avenues and streets . . . Dick Groeveveld wrote a number of stories, and the letters and words of his text constitute the whole visual architecture . . . [transforming the city] into a kind of three-dimensional book . . . where each bicyclist, by choosing their own path . . . makes a unique and personal reconstruction of the city. In the more recent Amsterdam and Karlsruhe versions . . . the size of each letter matches the size of the building at the location which it replaces.

THE VIRTUAL MUSEUM is a 3D computer generated museum containing an immaterial constellation of rooms and exhibits. Its apparatus is a round rotating platform on which is located a large video screen, computing equipment, and a chair on which the viewer sits. From the chair the viewer interactively controls his/her movement through THE VIRTUAL MUSEUM. Forwards and backwards movement of the chair causes forwards and backwards movement of the viewer's eye position in the museum space represented on the video screen. Turning the chair causes a rotation of the virtual image space and also a synchronous physical rotation of the platform itself, so that the viewer moves simultaneously in both the virtual and real environments . . . constituted by five rooms, all of which reproduce the architecture of the real room in which the installation is located. . . . Using alphabetic and textual forms, each room contains its own specific constellation of computer generated virtual exhibits.

> —from Jeffrey Shaw's "Modalities of Interactivity and Virtuality," artist's statement for a talk given at the Museum of Modern Art, "Technology in the '90s" series, 26 April 1993

What happens when we go to the museum?

This question creates a space within which there is a triple time: what happens to us as we go? what happens to the museum? what happens outside it and us?

What happens to the world, the work, and to us as we go?

Elsewhere nearby, a skim of water whispers on a mottled courtyard wall, a distant bomb thuds in a tower garage; here an earphone-clad herd turns on cue to a captioned wall and reads curatorial prose before shuffling past the paintings toward more text; within us "something that appears like nothing can take the place of something." Yet we begin to feel that this last "within us" somehow happens in the work of art, it becomes in Hélène Cixous's phrase the "betweenus we must take care to keep" (1991, 62).

What happens in the work of art is *the work of art.*

There exists a classicism, a formalism, and an interaction, each of which try to make this statement. The work, the world, the viewer, and the artist orbit like molecules around a thing. Jeffrey Shaw rightly identifies this with the Western "idolatry of... objective reality separated from the mind that creates it" (1993). We expect this will change with the future. Yet the problem with virtual reality (which is to say our imagination of the future) is that it attempts to reproduce the supposed, quite illusory, seamlessness of the aural, visual, and psychological world as thing. Yet what happens as we go happens at the interstices, at the place where the pure form of any thing is jostled by recollection, accident, or recognition and so slips into inscription i.e., beneath the virtual seat cushion where a jumble of spilled tokens, sesame seeds, abracadabra keys, and floating boxes of light transfigure us.

Shaw suggests that virtual reality makes "the real disappear within the deluge of hyper-real fabrications." But I do not think the real disappears. Instead, it never really appears except when it veers at the interstice of the "betweenus"; i.e., it never really appears but in what Shaw proposes as a "discourse in that fine zone between the virtual and actual." The viewer of the legible city or the virtual museum inscribes herself upon the gap. This is the interactive moment, what hyperfiction author Carolyn Guyer calls the "complex mixtures of polar impulse... the buzz-daze of space-time" (1993a). This is what happens as we go. At this interstice of Guyer's we encounter not the poles but, rather, the "third and infinite option [of] mingling, the coalescent, rhythmic ability to create nothing from anything." What the viewer writes is mingled history, history of the self as sign and desire, which is to say dance. She moves into and from meaning in a weave that both means and shifts meaning by her movement.

The interactive cinema artist Grahame Weinbren recognizes the mingled history of meaning-by-movement when he points toward "the misconception that *choice-making* is the primary token of interactivity, when *response* is a more germane concept" (1993a). Seeking a paradigm and practice for an interactive cinema in Freud's notion of condensation, Weinbren notes the buzz-daze of space-time that the dreamwork entails. I.e., the dreamwork is "not a narrative that unfolds in time—all the elements are simultaneously present... thus it is truly a narrative *without* specificity of sequence." The development of virtual reality is at present (and perhaps perpetually) stuck in specificity, awash in a tidal wave of infotainment investment, freed of the substance of capital and afraid of the infinite option of mingling. As a result virtual reality seems lost in a headlong, whitewater descent, floating through the deluge along the calm, Cartesian straits, propped above the flow of space-time upon a swollen raft of seamless interaction. In attempting a continually remapped seamlessness, virtual reality not only ties itself to specificity of sequence but also, at least at present, to a hopelessly mechanistic model of interface and (as) narrative in which the story space is, like that of the Legible City, one of successive disclosures of self-generating spaces. We are never in the water, never inscribing the flow even in our sailing.

Weinbren's cinema is built upon an understanding that inscription bubbles up from the level of system, i.e., "in the true interactive cinema, everything is in flux, subject to constant change... and the more sophisticated the system, the more fluid and wide-ranging the possibilities" (1993a). Interaction is thus seen as what Deleuze and Guattari term an eccentric science, "a hydraulic model... one of becoming and heterogeneity, as opposed to the stable, the current, and the constant": "It operates in an open space throughout which things-flows are distributed, rather than plotting out a closed space" (1983, 363).

Unlikely as it seems, a computer system, too, is an open space throughout which "things-flows" are distributed, a weave and a dance beneath. It is, as the Danish hypermedia theorist Peter Bøgh Andersen suggests, "nothing but a complex network of signs": "Everything in [it] from top to bottom, is used as sign... the programmer [does not]... create signs. What he does is rather to propose signs.... [P]rogramming in a semiotic perspective is not to instruct the machine to do something, [but] to use the machine to try to tell people something" (1994, forthcoming). From such a semiotic perspective an interactive installation from bottom to top is a dynamic dance and a polymorphic weave of its own becoming, what in another context I have called a structure for what does not yet exist. From the first, its network of signs creates itself as a space within which, also, there is at least a triple time: What happens to the data as it goes? what happens to its representation? what happens within what gives it expression?

At absolute bottom on the chip's surface, the Legible City is only an attempt to conserve energy and to dissipate heat as electrons try for the shortest cut along photo-engraved gold highways and back alleys across the city of the integrated circuit. Lacking abstraction, they know nothing about us or any city. Just above this, at the level of registers within the computer memory, is the coded matrix of abstract representation whose values, stored geometrically and understood algorithmically, propose a representation of what in this case Shaw describes as "a dimensional database 2.5 kilometers square." At a still higher level of abstraction, but on a paradoxically more representational plane, graphical algorithms attempt to shape another matrix, this one composed of elements of pixellated light, into a correspondent mapping on the screen. At yet a higher level of abstraction only the viewer can ferret out form and disclose it within the dance of time. Yet even here the system again asserts its uncertain claims to try to tell the viewer something, "a hidden meaning," offers Shaw, an "archaeological expedition [to where] these hidden stories are being told again" (1991). Shaw's attempt to "embody tentative structures of meaning" finally results in a discourse at the highest virtual level of the system, i.e., its surrounding environment, at the interstice of the so-called real world, the actual museum space in which we go.

At this last surface we try to tell what happens to us. Yet what we mean by "telling what happens to us" in this kind of formulation is inherently ambiguous. I.e., either we may think of it as trying to tell our own experience (where the emphasis is upon what happens), or we may think of it as our trying to make this experience communal (where the emphasis is upon us). Here then I would differ with Shaw's claim that the Legible City reveals its hidden meaning and, instead, suggest that what it reveals is rather its own, and our own, hiding of meaning. Weinbren speaks of this kind of revelation in the interactive cinema as the "epistemological state [of] a subjunctive relationship to the screen—*subjunctive* in the sense that the viewer is constantly aware that *things could have been otherwise*" (1993a). What we see when we encounter the Legible City or the Virtual Museum appears to be an ideogram: body and bike in ever unfolding but never closing alpha-architectural space, or chair circling ever a little more than arm's distance away from a flickering window into geometrical space. For museum goers who view the viewer on the bicycle or in the circling chair, these ideograms write themselves upon the actual museum space in an awareness that in space-time things always could have been otherwise. Meanwhile, from her subjunctive perspective upon the bike or chair, the viewer sees her own actions as writing themselves, i.e., as if they had meaning rather than hide meaning.

One genius of the high modernist moment lay in just this slippage from *had* to *hide,* which is to say in the substitution of the method of collage for the medium of the ideogram. Thus, Ezra Pound proposed that the ideogram—which has its own history printed, however opaquely, on its back—offered a fitting mode for an age that no longer could decipher a past it carried with it nonetheless. This was a mode in which Pound (and to a greater extent James Joyce) saw through in every sense, but especially the palimpsest. Hidden meanings avail themselves to whomsoever is clever at peeling away surface from surface. Gertrude Stein, who saw what the painters saw, knew differently, again in every sense. Hers is the method of mingling, the dance of history in which hidden meanings of the self veil themselves as tender buttons of sign and desire. Collage means on the whole as our eyes move over it but what makes its meaning in particular is not available.

The same is true of the Legible City. It too means on the whole, but what makes its meaning in particular is not available. On one level it is not available because it is, as we have seen, written in registers upon the chip and the matrix of light on the screen. On another level it is not available because it is beyond what the viewer of the viewer, the museum goer, can see—i.e., as Shaw explains: "The space of visualization was virtually located beyond the surface of the projection screen and thus outside the actual room where the bicyclist was situated" (1993). On still another level it is not available because we cannot see what the viewer brings to it. We see her on the whole as our eyes move over her, but what makes her meaning in particular is not available. This is to say that the viewer herself embodies an axis of symmetrical and interstitial relations. The city space fans open before her as a series of slippages and partial statements of an archaeological past (or, in the New York installation, a fictional future), which ever linger as after-images within which her movement creates new space. The city was here and left these traces, which she cycles through in every sense. To this vortex the viewer brings a fanning trail of after-images, cycles, and traces of her own, which drift behind her like the silky, smeared light of the milky way. From the vantage point of the museum goer (the viewer of the viewer) there are streets before you, there are galaxies behind you, each composed of light that illuminates no object.

As you view the viewer of the Legible City, the work itself begins to suggest the "Mannerist ambiguity" that Shaw envisions for the later Virtual Museum, in which "different orders of simulation are located in a meta-dimensional structure that mirrors a confluence of the real and the fictional" (1993). Yet this is theater as much as virtuality. The embedded confluence of the real and the fictional is neither in its conception nor its technology as much a brave new world as it might seem to those of us who do not recall interstitial Arden, the virtual forest of Titania and Oberon, or the technologized isle that once was Sycorax's. As theater, virtuality and interactivity enact nothing less than reading embodied. All theater, of course, is reading literally embodied. Meaning there is seen as interstitial and in movement, internalized and in triple time: What happens to us, what happens to the space we watch, what happens in time outside the space and us?

The promise of work like Shaw's, I think, resides in how its theater invites our interaction and inscription, how in the words of Prospero's epilogue it requires the "help of [our] good hands . . . else [its] project fails, which was to please." Unlike virtual reality, which seeks to recapitulate the high modernist moment in the seamless, knowable space of the programmed ideogram drawn in light, Shaw's works instead present virtuality as a collage of surface upon surface, knowable only in whole, which is to say knowable only in movement—whether of the eye of the viewer, of the museum goer, or the constant movement of the installation as at every level it inscribes itself. Yet Shaw's interaction somehow falls short of representing adequately the subjunctive epistemological state of seeing ourselves in triple time. What's missing from Shaw's work is what Weinbren identifies as the necessary "depiction of the type of transformation between one element and the next [which should] be incorporated alongside the transformations themselves."

I.e., it is not just the viewer nor history but also the computer screen that cycles (just as the city too cycles around it and the exhibition space cycles through its seasons within the museum space). Shaw's subject and modality as well as our shared interests in interaction invite attention to the fundamental nature of electronic media and how we inscribe ourselves within them as we go. Although Shaw's creation of "immaterial constellations" and "provisional" architectures, I think, place his work clearly in much the same realm as hypertext, yet neither of these works seems to have (or, more important, to induce in us) a conscious awareness of the transformation of narrative fundamental to hypertext. Elsewhere I have proposed, as something of a mantra, the following distinction: print stays itself; electronic text replaces itself (1992). The "radical transformation" of Weinbren's "'subjunctive' state of mind" is much closer to the hypertext's constant replacement:

> In normal cinematic circumstances, the weight of an event is given largely by its context: now, with sequence under the thumb of the viewer, this weight can alter or even dissolve. And in those circumstances the viewer's relationship to, his understanding of, the events of the narrative can undergo a radical transformation, based entirely on the knowledge that *things could have been different.*

With electronic text and graphics as well as with video and interactive cinema we are always painting, each screen unreasonably washing away what was and replacing it with itself. Thus, even a viewing of a static screen becomes a constant pentimento, a collage of one-to-one mappings that we can understand in whole. Yet, however much it seems so, meaning in particular is not available within this constantly replaced surface. Instead, as in Weinbren's interactive cinema, meaning is "grounded in the viewer's continual knowledge that what is on the screen is a result of her interactions—inaction, naturally, being as much a determining factor as action" (1993a).

Hypertext as an art form concerns itself with constant reconfiguration and so is a true electronic medium. Hypertext is before anything else a visual form, a complex network of signs that presents texts and images in an order that the artist has shaped but which the viewer chooses and reshapes. Truly interactive, constructive hypertext proposes choices in such a way that what happens in the viewing causes both the shaped form and the available choices to change. The interaction that hypertext aspires to is its own reshaping, an interaction some may see as its own overturning.

Hypertext vindicates the word as visual image and reclaims its place in the full sensorium. It is the revenge of the word on television. Offering both a mode for the interstitial narrative and a medium of interstitial imagination, hypertext writes in the gap between the lines and the senses alike. To use Weinbren's terms, hypertext incorporates the type of transformation between one element and the next alongside the transformations themselves. Thus, it is likely that as hypertext inevitably merges with virtual reality, virtual reality will emerge from its current techno-onanism. At present the urgent, voyeuristic gaze of virtual reality—because it is as yet so visually primitive—must enlist and subvert what we might call the Cubist, multiple, or provisional vision of the free subject and appropriate it to filling in needlessly the seamless body of objectified representation. All the hard-won vision of the twentieth century is to be surrendered to wire-frame realism in the twenty-first. This subversion of the viewer is what makes the real disappear into Shaw's hyperfabrication.

Yet the problem we face is not to fill in a world but, rather, to fall into it. We give in to time; we give way to time; we give in with time. The world we grow into is full, or so, increasingly, we think. Out of time we seem (literally) to be running out of space. Everything is (over)written, and it seems, in Luce Irigaray's phrase, "dangerous to believe" even in nowhere (the Greek, *ou-topos,* or utopia). Thus, the question of gaps or interstices is problematic. What do we call a gap that is full, a question Irigaray (1993) calls us to? What sort of island (what utopia or dystopia) could we possibly establish in the turbulence of William Gibson's "bright lattices of logic... the consensual hallucination [of] the matrix" (1984, 5)?

Against the word and world fully mapped as logos, Deleuze and Guattari propose that we write ourselves in the gap of *nomos,* the nomadic (1983b). They pose wandering against the word, being-for space against being-in space. We are in the water, inscribing and inscribed by the flow in our sailing. We write ourselves in oscillation between the smooth space of being for-time (what happens to us as we go as well as what happens to the space in which we do so) and the striated space of in-time (what happens outside the space and us). Interactive electronic art seems to offer a paradigm for just such an oscillation, a constant becomingness as a way out of what we are in, a way in where we are put out. Interactive art gives way to giving ways: *Things could have been different.*

I speak about the becomingness of hypertext in terms of contour. Contours are the sensual whole that we move over: transitory, evocable, multiple, and generative structures that make up our experience of interactive arts. We most often perceive contour in the sensuality of constantly reinscribed, impinging surfaces of word, image, and perceived shape. "If there is a name *surface,*" asks the poet Erin Mouré, "then what else is there? is what is 'different' from the surface *depth* or is it *another surface?*" (1988, 90). Contours are the shape of what we think we see as we see it but that we know we have seen only after we move over them, and new contours of our own shape themselves over what they have left us. They are, in short, what happens as we go, the essential communication between the artist who gave way and the viewer who now gives ways to see.

It is not entirely clear whether Shaw is aware of the extent to which in the Virtual Museum we choose *to see,* not *what* to see. What we do see by his accounting are "five rooms all of which reproduce the architecture of the real room in which the installation is located." The first room is an echoic and inscripted space that, in Shaw's words "shows a representation of the installation itself, with its platform, computer, video screen and chair." But, we note (tellingly), not the viewer. What is echoed and inscribed is the action of "trying to tell." Yet the telling that inscribes itself upon the space is a complex network of particular meaning not available to the artist but only to the body in history. It is an actual dance, in which viewers, caught up as Shaw says in "a world that is becoming increasingly museified" and threatened with "premature conservation," can move through "a virtual museum architecture as provisional as the culture it embodies" (1993).

We might, rather, say that it is the viewer by her interaction who embodies a provisional culture as much as the architecture. What happens as she goes is that she inscribes herself upon the culture as a presence necessary to its encoding. The viewer leaves Shaw's first room aware of what it has not represented, i.e., Weinbren's "determining factor" that her "continual knowledge that what is on the screen is a result of her interactions [or] inaction." The anthropologist Michael Herzfeld suggests that "between social and monumental time lies a discursive chasm, separating popular from official understandings of history": "Social time is the grist of everyday experience... in which events cannot be predicted but... every effort can be made to influence them." At the interstice, in that fine zone between the virtual and actual, the viewer (to use Shaw's verb) relieves mind and history alike from museification. This is to say that at the interstice, social and monumental time mix. Social time, according to Herzfeld, "gives events their reality, because it encounters each as one of a kind": "Monumental time, by contrast... attempt[s] to place time and history outside the bounds of lived human experience" (1991, 7).

The impinging surfaces of the rooms in Shaw's Virtual Museum mean to present successive contours of seeing itself. The sailing, shifting letters of the last room weave light from light. Their dance induces the viewer to see that signs map themselves, if not deceptively, then abstractly. What moves and reads as *A*, *2*, and *Z* the viewer discovers in the buzz-daze as *red*, *green*, and *blue*. From a semiotic perspective Shaw's system tries to tell that the specificity of this room, unlike its antechambers, is that of a "wholly computer generated environment." It is a culmination, however, not a closure. *Closure*, to the extent that it is any longer a viable term in interactive arts, has been described by Andrew Sussman as "the completion of self by the reader" (1988). Or as Weinbren puts it, "a viewer uses an interactive work until she is finished with it... and not until it is finished with her" (1993a). Only the personal history of the viewer marks the small scale events of Shaw's "continuously changing coloration... generated by the radiance of these moving signs" (1993). She may contain them but does not close them off. The essential communication here is between given ways and ways to see. Shaped by these spaces, we choose to see, not what to see.

Interaction replaces (reinscribes) time and history within lived human experience. The archaeologist Rosemary Joyce argues that, by placing the authority for a particular view of the past in a self-contained materialization (a sculpted monument in Maya Classic), "monumental time can be said to assert an independence of the personal history and small-scale events that make up the life of people" (1993, 4). Lived human experience is relieved. In much the same way, by placing the authority for any view (including the view of the viewer) in what happens as we go, the interactive installation asserts its dependence upon personal history and small-scale events.

Social time emerges in relief from monumental time. What Herzfeld calls the unpredictable events of everyday experience survive every effort made to influence them. Since the meaning of what the viewer chooses to see is not available—only that she does see or that we see her seeing—the interactive work asserts the independence of social time inscribed in and beyond the monumental time of the museum.

Shaw's installations may likewise point us toward a way out of commodified time, all we have left of monumental time in our age. Like monumental time, commodified time also attempts to place time and history outside the bounds of lived human experience. "Is it possible to imagine an aesthetics that does not *engage,* that removes itself from History and even from the Market?" asks Hakim Bey, "or at least [one]... which wants to replace representation with presence?": "How does presence make itself felt even in (or through) representation?" (1991, 130). The Legible City evokes a continuum from the high mechanical age of the ideogrammatic bicycle to its degradation into icon in the logotype of ET's bike against the sky. Similarly, the chair and screen of the Virtual Museum evoke the Mad Hatter's teacups of the Disney amusement park ride, the TV chairs of airports and bus stations, and the Star Trek deck in turn. These interactive works are self-consciously art objects, containing virtual art objects, engaging human viewers whose interactive movements become objects of art, whose body means by movement. It may be that, as we view the viewer cycling through video projection or swiveling enthroned before her screen, we relieve and disengage both her and us from commodified time and inscribe her presence in social time.

Presence makes itself felt in or through the body. Since the place of the affirmation of social time by the monumental is the body, interaction finally takes place within the dance of space-time and the weave of human evolution. "Inside and outside the body is one," says hyperfiction writer Martha Petry, "geographic, architectural, with permeable boundaries... collective and communal": "This continual movement is what forms the narrative... of what we do and how we make what it is we own" (1993, 3). It is a reconfiguration, however, not a closure. It is, in short, what happens to us as we go.

> [A] primary concern has been to retain the articulation of time. Without it, we quickly descend into the pit of so-called "multimedia," with its scenes of unpleasant "buttons," "hot spots," and "menus," [which] leaves no room for the possibility of a loss of self, of desire in relation to the unfolding... drama. The ideal machine for an interactive cinema has a screen that is continuously responsive... through which the individual viewer can always make an impact on the sound/picture complex. The central problem in creating a work for this medium becomes that of integrating continuity and interruption—balancing the continuity of the film against the interpretation caused by the viewer's responses to it.... The primary source for Sonata is Tolstoy's short story, "The Kreutzer Sonata"—a study of jealousy and distrust, which culminates in a man's explosion into murderous rage.... [T]o examine extremes of emotion... it is possible to see alternate views of the same situations... caught up in the drama but at the same time analytically removed from it. Further the availability of a variety of ways of engaging the same story highlights the differences between memory and reality, and it reveals the selectivity of memory.... The piece incorporates... the haunting dream-image... ascribed by Freud to the... Wolf Man [by] the interactive strategy [of] a kind of peeling away: at viewer inter-

ruption, the current layer is removed, revealing its sources in the viewer's psyche.

The Judith [and Holofernes] story is developed visually through its multiple portrayals, and a narration based on [its retellings] The viewer navigates . . . a unique interface [representing] temporal direction. Thus for example, pointing at . . . the left side of the screen changes the direction of time. However when we go back through events, as in remembering or retelling them, the events are necessarily seen in a different light. Memory is not pure factual description—it always incorporates desire, explanation, justification; unlike computers, we remember what is important to us. So when the viewer actually activates "reverse time," a new view of the events is revealed.

—from "Pointing at an Interactive Cinema," Grahame Weinbren's artist statement regarding *Sonata*, 1993

On Vassar Farm where during the time I began to think about this essay I walk each morning and again at twilight the mown margins along the road are woven under with buttercup, thistle, clover, daisy, dutchman's breeches, and several kinds of flowering grasses unknown to me, all of which edge further off beyond the road into high and duller grass, their hues flaring, dotted, clustered in perceptible patterns until absolute thickets of undergrowth contour hills of impenetrable tangle, bramble, fallen boughs, and thick trees, much of it threaded through with grape vine and patched with wild rose, which sometimes weaves or cascades through deadwood—the whole deciduous jungle so thickly perfumed at nightfall the air has its fragrance yet through which deer easily pass at will and rabbits disappear like robots. Here also meadows of hay yellow between cuttings, dark green bamboo fringes jade scummed wetland, furrow leads to swale, swale gives way to depression, depression to upland rise, green to darker green. Intervals along the road disclose a softly worn distant shoulder of a ridge of Catskills, often misty or heat shimmering, rising in ancient contour and then hidden again by another turn in the road.

It is, whether in language as dense or less dense than this or in virtual systems of varying densities, a representable space. I believe the actual space at hand is one I create. This is the nature of continuous representation. My belief that I create the space I walk through is not merely an artifact of language. A woman walks toward me and passes, for all I know wordlessly, yet I believe that she too creates the space she walks through and that it is somehow different from mine. That even a wordless walker creates such a space is less and less something that admits to serious challenge on philosophical, perceptual, aesthetic, or technological terms. What we are apt, instead, to challenge on those terms is any claim for a fundamental difference between the continuous representations of the composing man and the unknown woman. What is under dispute is not so much the nature of choice but, rather, nature and choice.

We believe that each chooses what she sees constrained by philosophical, perceptual, aesthetic, or technological factors and that the continuous representations each one creates result from what we might call the subtractive effect of these constraints. This is the search-space problem, buzz-daze, the frame. I see the willow; she sees the loosestrife.

"An interactive cinema," according to Weinbren, "intermixing and interweaving multiple narrative streams can create a metanarrative stream . . . greater than its component parts, if the subject matter . . . is a match for the potential of the medium": "The ideal, utopian subject is that of the human mind in operation" (1993a). I see the willow; she sees the loosestrife.

When she tells me what she sees, I change my representation accordingly, constrained—or say what I see and, in so saying, likewise change my representation, accordingly constraining. (What we see is, as Weinbren says, "grounded in the viewer's continual knowledge that what is . . . is a result of her interactions—inaction, naturally, being as much a determining factor as action" [1993a].) We walk on without speaking.

"People are multi-tasking beings, parallel processors," suggests Weinbren, "they can whistle and daydream while working, phantasize while having sex, speak the English translation while listening to the German, and so on, switch from one mental activity—one state, one condition—to another instantly and without effort" (1993b).

The dispersion of wildflowers, etc. is, if not knowable, at least representable, projectable, in say a scatter diagram or computer rendering, though not at the level of continuous representation. I do not know what she is thinking, although I do know what she is thinking about. No, I do not know what she is thinking about, although I do know she is thinking. No. To repeat Weinbren's dictum, the dreamwork is "not a narrative that unfolds in time—all the elements are simultaneously present . . . thus it is truly a narrative *without* specificity of sequence" (1993a).

One could (I could), of course, believe she created the loosestrife (since I had not seen it before her seeing), but that would require extra-world actors: princess, goddess, a woman walking.

On the internet mail list in 1993 TNC (TechNoCulture) Don Byrd, poet and professor of English at SUNY at Albany, "plead[ed] for attention to Spencer-Brown's *Laws of Form*":

> Here is a logic in which the logos is not "always already" a product of a mysterious original space (i.e., the object of the successful poststructuralist critique). Forms are rather generated in real time by virtue of living organisms cognizing their being. Forms are time forms, not space forms . . . we arrive at a time when we have to learn to think a formality which is not present in space but in the temporal duration.

Narrative as metric.

Deer easily pass at will and rabbits disappear like robots.

It is as Calvino imagined:

> Naturally, we were all there—*old Qfwfq said*—where else could we have been? Nobody knew then that there could be space. Or time either: what use did we have for time, packed in there like sardines?
>
> I say "packed like sardines" using a literary image: in reality there wasn't even space to pack us into. Every point of us coincided with every point of each of the others in a single point, which was where we all were. (1965, 13)

The interdeterminability of points of perception argues against a virtual reality that depends upon successive disclosures of self-generating spaces. VR requires the interstitial drama of Weinbren's "depiction of the type of transformation between one element and the next . . . incorporated alongside the transformations themselves." A woman walks toward me and passes, for all I know wordlessly, yet I believe that she too creates the space she walks through and that it is somehow different from mine. (I.e., there is no point from which to see, even in 3-space, each new point in its own perspectives.)

"The most interesting possibility for me... based on my interest in the multiplicity of narrative," writes Weinbren, "is that the narrative lines continue until one several or all of them ended": "The viewer navigates from one current to an adjacent one in a constantly flowing river, crossing between streams of a story at moments of similarity or juncture" (1993a).

Contour is one expression of the perceptible form of a constantly changing text, made by any of its readers or writers at a given point in its reading or writing. Its constituent elements include the current state of the text at hand, the perceived intentions and interactions of previous writers and readers that led to the text at hand, and those interactions with the text that the current reader or writer sees as leading from it. Contours are represented by the current reader or writer as a narrative. They are communicated in a set of operations upon the current text that have the effect of transforming that text. Contours are discovered sensually, and most often they are read in the visual form of the verbal, graphical, or moving text. These visual forms may include the apparent content of the text at hand; its explicit and available design; or implicit and dynamic designs that the current reader or writer perceives either as patterns, juxtapositions, or recurrences within the text or as abstractions situated outside the text.

An eccentric science: "Or they might rather be thought of as potential narrative streams," continues Weinbren, "elements themselves unformed or chaotic, but taking form as they intersect, gaining meaning in relation to each other" (1993a).

On the day after I gave the talk entitled "What Happens as We Go?" at the Museum of Modern Art (MOMA) (26 April 1993) the cyberartist (*Agrippa*) Dennis Ashbaugh said: "My two fiber optic guys walked out halfway through your talk. They said you don't know shit about computers or VR." Two days later he left a long message on my machine asking me to call. We didn't connect until a week later when he called to say that his fiber optic guys had called back within two days to say that the talk had haunted them. "It haunted me too," he said. "Some of what you said comes back after a while and it sticks. It's weird you know. Do you know what I mean?"

In *Sonata* Weinbren juxtaposes the stories of Podsnyshev (Tolstoy's *Kreutzer Sonata* antihero wife stabber) and Judith (the Apocryphal heroine who decapitated the enemy general Holofernes). Both narratives progress, but it is at their interstices, where the viewer crosses from one to the other, that they come into focus and take on meaning. A viewer will see an episode of one or the other narrative but not both, and their forms are similar enough that their plot movement can be seen as concurrent. Each of the two killers, Podsnyshev and Judith, is reflected in the light of the other, since each emerges out of the story context of the other. And thus an act of interpretation is forced on the viewer: the morality of each character comes into question when they are placed in relation to one another, especially because the former is conventionally thought of as Evil and the latter as Good.

(The whole of the preceding paragraph—with the exceptions of my substitution of "Weinbren" for "I" in the first sentence and "interstices" for "connections" in the second—was taken verbatim from Weinbren's 1993 unpublished essay. Weinbren also was present at the MOMA talk; he had just flown in from the West Coast and said it was hard to follow the talk because it was dense but "dazzling." Laurie Anderson was also there, and, though I could not see clearly—and though, also, she afterward grasped my hand and smiled while saying she enjoyed the talk—I think she dozed in the darkened auditorium. About a month later I spent nearly an hour with *Sonata* at Weinbren's studio, and my responses were very close to those I describe in using his paragraph. The original talk, of course, was only about Jeffrey Shaw's work and followed Shaw's own talk and demonstration. The kind of self-referential text is characteristically postmodern, I suppose, but it brings to mind Italo Calvino's extraordinary 1967 essay, "Cybernetics and Ghosts" (collected in Calvino 1982) (a year after Glenn Gould's claim in "The Prospects of Recording" that "indeed all the music that has ever been can now become a background against which the impulse to make listener-supplied connections is the new foreground" [1966, 17]).

Calvino rightly sees postmodernist vision as increasingly discrete rather than continuous. Electronic brains provide a convincing model of the mind as network (he is reading Shannon and others, seems aware of early AI work in chess). "In the place of the ever-changing cloud that we carried in our heads... we now feel the rapid passage of signals" (1982, 8).

(Cf. "Quickness" in Calvino's *Six Memos for the Next Millenium*: "In an age when other fantastically speedy, widespread media are triumphing and running the risk of flattening all communication onto a single, homogeneous surface, the function of literature is communication between things that are different simply because they are different, not blunting but even sharpening the differences between them, following the true bent of the written language" [1988, 45].)

He suggests that to say that "what is finite and... calculable is superseding the indeterminateness of ideas" is false because, the more we can split the world (Zeno), the greater the expanse of what we do not know. "The process going on today is the triumph of discontinuity, divisibility, and combination over all that is in flux, or a series of minute nuances following one upon the other" (1988, 9). He then cites Watson and Crick, Greimas's actants, Oulipo, to claim that in history, biology, and especially structural linguistics/semiotics "the endless variety of living forms can be reduced to the combination of certain finite quantities" and that "mankind is beginning to understand how to dismantle and reassemble the most complex and unpredictable of all its machines: language" (12).

Then he begins to speculate "a writing machine that would bring to the page all those things we are accustomed to consider as the most jealously guarded attributes of our... life" and argues that, given enough rules, such a machine would be a classicist (literary avant-garde is "still an entirely lyric instrument, serving a typical human need: the production of disorder"), but that "nothing prevents us from foreseeing a literature machine that... feels unsatisfied with its own traditionalism and starts to propose new ways of writing" (1988, 13).

(Slyly, Calvino proposes that it could also be programmed with Marxist and sociological determinisms so as to keep litcrit types and journalists in work.)

(In an exchange of e-mail that followed my posting of these notes about Calvino on TNC—as part of an extended exchange with Byrd about hypertext narrative, proprioception, time forms, and the characteristically postmodern self-referential text—David Porush steered me toward his own, earlier discussion of the Calvino texts, and "Cybernetics and Ghosts" in particular, in which he argues that, "though later in the essay Calvino denies it, it is hard to believe that he isn't being slightly whimsical": "Yet even if we take him at face value as he asks us to, it is clear that Calvino's version of the author-machine is a very special sort of cybernetic device, one which at least has access to the unconscious and the mysterious" [1989, 373].)

The speculation about machine literature satisfies Calvino's own uneasiness about what on TNC we were calling "an author in ROM," i.e., story generation, though obviously I think it also bears upon interactive story generation: "Various aesthetic theories maintained that poetry was a matter of inspiration descending from I know not what lofty place, or welling up from I know not what great depths... But in these theories there always remained a void that no one knew how to fill, a zone of darkness between cause and effect: how does one arrive at the written page?" (Calvino 1988, 14).

(What happens to the world, the work, and to us as we go?)

And leads to an admission that "Writers... are already writing machines, at least they are when all goes well..." and to an assertion that "once we have dismantled and reassembled the process of literary composition, the decisive moment of literary life will be that of reading." This prepares the way for a reconsideration of the storytellers tasks from a reader's perspective. Fable (story) gives way to myth (the unsayable), and it is the real person of the reader (as writer) who actualizes it. "The work will continue to be born, to be judged, to be destroyed or constantly renewed on contact with the eye of the reader" (Calvino 1988, 16).

In a remarkable paragraph Calvino spatializes writing: "Literature will continue to be a 'place' of privilege within the human consciousness, a way of exercising the potentialities contained in the system of signs belonging to all societies at all times." Having done so, he is ready to dispose of "the author: that anachronistic personage, the bearer of messages, the director of consciences, the giver of lectures to cultural bodies." And once disposed to herald the second coming: "And so the author vanishes—that spoiled child of ignorance—to give place to a more thoughtful person, a person who will know that the author is a machine, and who will know how this machine works" (1988, 16).

Here at last (writes Weinbren) the Interactive Cinema can become a Vessel of Desire (his capitalizations) in its insistence on the viewer's role in the making of meaning. Without an act of interpretation, the stories are raw and problematic, but, when clashed together at the points of interaction, a role of the Judge (his capitalization) is forced on the viewer.

Or as, also writing of desire, I wrote in the essay (itself first a talk and alluding to Laurie Anderson) that follows in the next chapter: Our desire is a criticism that lapses before the form and so won't let form return to transparency, a criticism that rather than stand still, constantly falls down dead center in the battle over who shall control the content.

> I never want this talk about desire to stop.
> I never want this talk. About desire to stop.
> I never want this. Talk. About desire to stop.
> I never want. This talk about desire to stop.
> I never want this. Talk about desire to. Stop.

Intervals along the road disclose a softly worn distant shoulder of a ridge of Catskills, often misty or heat shimmering, rising in ancient contour and then hidden again by another turn in the road. In the fourth year we both, it seemed, came to the conclusion that we could no longer understand each other. We no longer tried to bring any dispute to a conclusion. We invariably kept to our own opinions about even the most trivial questions, but especially about the children. As I now recall, the views that I maintained were not at all so dear to me that I could not have given them up, but she was of the opposite opinion, and to yield meant yielding to her, and that I could not do. It was the same with her.... When we were alone we were doomed almost to silence or to conversations such as I am convinced animals can carry on with one another. I dreamed that it was night and I was lying in my bed. Suddenly, the window opened of its own accord, and I was terrified to see that some white wolves were sitting on the big walnut tree in front of the window. There were six or seven of them. The wolves were quite white and looked more like foxes or sheepdogs, for they had big tails like foxes, and they had their ears pricked like dogs when they pay attention to something. In great terror, evidently of being eaten up by the wolves, I screamed and woke up.

The Momentary Advantage of Our Awkwardness

An inverted essay in two parts, the second first, for reasons that may become clear in reading the first. "We are struggling against all forms of other to be able to live," writes Luce Irigaray, "and we are still subject to conditional social rules that we confuse with freedom" (1993, 28).

Hypertextual Rhythms (Part 2)

In "the war between passion and technology" (as Porush calls it [McCaffery 1992]) we can take no sides. Passion or technology—each is us; we are the sides. So, we ought to approach as Cixous says Clarice Lispector does, toward the "betweenus we must take care to keep" (1991). There we surrender to ourselves this battleground, which we endlessly invent and represent.

In the course of our endless representation, aflow in the heavy water of what in another time I called the aleatory convergence, we are, we think, aware of the merging of somethings into Something (Sum Thing = Sun King). Hypermedia, multiple fiction, virtual reality, autopoesis, semantic space, simstim, cyberpunk, cyborg, grundge, rave, sweet honey in the rock and roll—it matters not what we call it, since our children's children will have been (and are) renaming it from moment to moment as by then they will have been used to (used for), our ancient children in the centuries after the apocalypse of the name.

We give in to time, we give way to time, we give in with time. I mean here an *a-polemikos,* a language of surrender, of declination, of rhythm, of flow. How can we win by surrender (or why win at all?)[1] is what I will ask you not merely to think about but also to act upon. "Give it up," as Arsenio says. This discussion is addressed to you, writer who was and writer who will be, graduate student and ancient child, passion artist and devotee of the electro/ecstatic. It means to beg from you a criticism that won't force apart the poles of passion and technology (these thighs) into *passé-systematique,* the past-systematic tense, but, rather, one that in giving way will give ways.

Our desire is a criticism that lapses before the form and so won't let form return to transparency; a criticism that rather than stand still, is always falling down dead center in the battle over who shall control the content.[2] We need to surrender control and in that constant declination continually render control meaningless. We need to be content and, in so being, become the content of our own passionate technology. Even now (as always) the epic struggle within the two-as-one will begin again, within Circe and Odysseus, sleep and waking, form and function, style and content, technology and passion. It is a struggle played out diagramatically, waltz steps stenciled upon strata of dream & waking across the parquet floor of the proprietary ROM chip; silicon-smooth space graven by images of the icy striations of software keys (quays) running with once boiling gold. At every level of the system—apparent narrative to interface, object to source, lexia to link, box to line, the writer who was and the writer who will be, Inferno, Purgatorio, Paradisio—the war goes on for the ownership of dreams.

Recalling the Sicilian storyteller's formula, "lu cuntu nun meti tempu" (which Calvino [1988] translates as "time takes no time in a story"), let me tell you a fairy tale from the *New York Times* about a promised land, a golden valley near a vast sea, where men (only men) live and work in towers of sand itself, a shiny city built of brick, glass, and gilt silicon:

"Gaston Bastiaens, who left Philips to head the Apple PIE division last summer... was not available for comment, but several people inside and outside the company said he has argued for a strategy similar to Philips and Nintendo's, under which the computer maker would largely control the content for Newtons" (Markoff 1992).

What content will we be content to surrender to le Roi du Pi(e)? How long can the apple fall in an emptied world?

Lu cuntu nun meti tempu. Let me show you seven citations in a row from Cixous (1991):

What is open is time: not to absorb the thing, the other, but to let the thing present itself. (63)

Metaphor? Yes. No. If everything is metaphor, then nothing is metaphor. (50)

medium my body, rhythmic my writing (53)

in order for the third body to be written, the exterior must enter and the interior must open out. (54)

the mode of passivity is our way—really an active way—of getting to know things by letting ourselves be known by them. You don't seek to master... [b]ut rather to transmit: to make things loved by making them known. (57)

(If you think there were only five, you have a good ear. If you think there were more than seven, you have a better.)

Lu cuntu nun meti tempu. This discussion is part 2 of something neither you nor I never heard (the negative doubled because we, of course, hear at least its echo now), a virtual talk solicited for a "virtual seminar" on the "bioapparatus" sponsored by The Banff Centre in Alberta, Canada, in the fall of 1991. In their invitation the seminar directors, Nell Tenhaaf and Catherine Richards, wrote:

> We constructed a residency on the bioapparatus... to look at the technological apparatus in its intimacy with the body, examining the history of this interrelationship and its sociocultural implications... [exploring] questions of the integrity of the body and of subjectivity. The apparatus... is itself a perceptual model, a reflection of social and cultural values systems, of desires.... Virtual Reality... is of particular interest... [as] a symptom as well as an instrument of the reordering of perception and... power relations.

(Most likely their formula means to defer having *tutoyer la cyborg.* After the seminar was done they put it in a book, which I too will almost surely give up and do with these two parts: perhaps calling one part passion the other technology, a recollection of a residence where I never lived.)

It was there I first spoke of the momentary advantage of our awkwardness: the time when the improbable mixed light of dawn isolates a half-illuminated world into angles and instances, the shapes we project upon its shadows in the instant before the bright blue or inconsolably gray closure that mark the resumption and progression of the seamless historical day.

In the transductive moment of the shift from one milieu to another, say Deleuze and Guattari, the milieus "are essentially communicating... open to chaos, which threatens them with exhaustion or intrusion" and to which "rhythm is the melieus' answer" (1983, 313). Dawn or twilight, hypertext fictions in the late age of print exist in a transductive moment that offers the momentary advantage of its own awkwardness. We are thus allowed to see each melieu in the moment before it assumes its seamlessness, at a time when they are essentially communicating.

Lately I have been writing, much too obscurely, about the simple idea of contours in hypertext (Joyce 1992). Contours are the shape of what we think we see as we see it but that we know we have seen only after they are gone and new contours of our own shape themselves over the virtual armature, the liminal form, the retinal photogene (after-image) of what they have left us. They are, in short, the essential communication between the writer who was and the writer who will be.

As rhythmic occurrences, they are products of what Terry Harpold terms the "pure spacing... upon which a narrative depends in order to anchor meaning among the sliding tokens of the telling" (1994, in press). Like a child in a fairy tale (the telling or the tale, the technology or the passion, it does not matter); like a child at Fujisney World, the fantasy island of the coming virtuality (those who do not understand the past are condemned to repeat sitcoms); like a child reciting before adults (mother&father, passion&technology) before you today, I beg you (I do!), "Please don't let it go away!" I surrender to the embrace only you can grant and only in the "betweenus," the passing distance that links us only as you (O mommy! please don't) go away.

What happens in this begging, this declination, is that we make time, literally create it out of (o blasphemer!) nothing. Nothing, i.e., but the sliding tokens, the rhythmic occurrences, the time of the repeated telling. "Not versions," as I wrote in *afternoon* (1990), "but the story itself in long lines." What would a criticism look like that does this? a criticism flowing without versions in one long line? a contour?

Lu cuntu nun meti tempu. Let me show you two citations in a row from Guyer (1992) (would she also read these? had she? I didn't know as I wrote. We all cite one another constantly, both making and marking time, surrendering to the others' voices):

> in that rhythmic sense of ebb and flow, of multi-
> directional
> change,... events... disappear before they're quite
> intelligible but
> somehow come to mean something...
>
> The thing about hyperfictions is that, for art, they tend to be extremely lifelike. They move and shift, allowing everything, and so allow only that we find our own perspective. They are so multiple they reveal what is individual, ourselves, readers of our own story.

Lu cuntu nun meti tempu. The problem that led me to consider the contour was this: we claim for constructive hypertext a surrender of writer to reader (the writer who will be), yet, if this surrender is not to sunder, it must be reciprocal. What I will write of (and *with* and *in* and *after*) what she writes will have to recall what she has written in such a way that the third self—you who writes after us—allow yourself to find me finding her first. Or else what either of us have done will be lost to you, and the space of the story will not be filled with time. Conversely, filled with time, the contour insists that versions are the story itself—what else could they be?

In the earlier part of this discussion no one ever heard I tried to situate the problem with virtual reality as it is presently constituted (presently situated, presently instantiated), in its attempt to reproduce the seamlessness of the aural, visual, and psychological world. Like the King of Pies, it wants to control Newton. The rupture-rapture-suture-future of VR as it stands means to be meaningful (to win the war of technology and passion) when it should be eventful or empty (and so fall down and surrender). "Inside and outside the body is one," says hyperfiction writer Martha Petry, "geographic, architectural, with permeable boundaries... collective and communal (much like storyspace's own). This continual movement is what forms the narrative... of what we do and how we make what it is we own" (1993, 3).

What it is we own: not a version but the whole. "Within the analytic tradition that parses complex flow as combinations of separate factors," N. Katherine Hayles observes, "it is difficult to think complexity": "Practicioners forget that in reality there is always only the interactive environment as a whole" (1992, 21).

In the second part of this discussion I will have tried to situate the problem of defining a fictional form within a critical discourse too much used to establishing versions. George Landow wonders why so many of today's critical theorists love to call themselves Luddites and thus "marginalize technology, which like poetry and political action, is a production of society and individual imagination," a condition that he sees as romanticizing "an unwillingness to perceive actual conditions of their own production" (1992, 168). What could that be but insistence on versions?

Stuart Moulthrop imagines a "reflexive critique . . . a deconstructive hypertext which would remind us that any system, even or especially one that advertises its own contingency, can have its authority called into question" (1991, 296). I have in mind a criticism without versions, one that redeems the method of Zeno, each new cut making space for time, each calling authority itself in question. Redeeming time, this criticism would surrender to Willie Wonka's physics of "So much time, so little to do" rather than the Disneyite dictum of a small world afterall. Take your time; there is so little to do. We can afford to wait for you to wait for the technology to settle into conventions. By that time the surrender will be lost, and someone will have to have won.

Perhaps you think this the same old song, another Lacanian, Deleuzian, Cixousian, utopian, wild again, whimpering simpering child again . . .

Lu cuntu nun meti tempu. "Did he say anything about fiction?" I give up. Will you?

The Momentary Advantage of Our Awkwardness (Part 1)

This curious formation, the bio/aparrat/us shares with the current state of representation in virtual reality, its intended target, the momentary advantage of its own awkwardness. We are thus allowed to see each in the moment before it assumes its seamlessness, as, e.g., the awkward and improbable mixed light of dawn isolates a half-illuminated world into its angles and instances, the shapes we project upon its shadows with our light-drunk eyes, before the bright closure of relentless blue, the inconsolable expanse of gray, or the anthropomorphic and comic clouds that mark the resumption and progression of the seamless historical day.

The first bioapparatus was the word: once outside the mouth metric, then formulaic; once imprinted, finally electric. Only in the late age of print can we again see the word in its awkward freshness and appreciate that it too was no more than apparatus (apparent to us) (a parent to us). "What has been the biggest change for you?" someone asked one of the VR pioneers. "I never thought I'd be used to flying," said the fly boy, happily outside history, not knowing Dante (or Beatrice before both fly boys). In the advent of the word, also, flying was an everyday thing.

The flaw in this most recent avatar of the bioapparatus is its projection upon a Cartesian and geometrical base. In VR the search for the coin of self beneath the virtual seat cushion is not metaphoric but, instead, proximate and actual. What's missing is the interstitial link, the constant mutability and transport that the word, now electrified (before VR was TVA), is not used to (for, despite its hegemony, it is not yet conscious; rather, it is we who are used to this in it) but used for. "The work of the morning is methodology. How to use yourself and on what. That is my profession. I am an archaeologist of morning," said Charles Olson (1974, 40). Apparatus: *ad/parare:* make ready. John the Baptist, light-drunk archaeologist. The interstitial link is Alice's archaeological looking glass.

What word shall we fly out on this morning, whence slip through the interstices?

Solitude "is perhaps the only word that has no meaning," writes Kristeva. "Without other, without guidepost, it cannot bear the difference that, alone, discriminates and makes sense" (1991, 12). She the apparatchik, like Ateh before her, protectress of the legendary Khazar dream hunters (see Pavič 1988).

Must there always be a word for it? scoffs the VR flyboy ("Why read when you can see, why see when you can experience?"). In the solitude of the bioapparatus VR is nothing less than an awkward instance of the new writing. Against solitude, only the word: once electric, then embodied, topographic, proprioceptive. In this case the word as interstitial link, what elsewhere I have termed the constructive hypertext (versions of what it is becoming, a structure for what does not yet exist) or the virtual seat cushion (a jumble of spilled tokens, sesame seeds, abracadabra apparatus, salt), which taken together inscribes with and upon our bodies the multiple fiction, the momentary advantage of its own awkwardness. (Virtual? Reality.) Against solitude let us seat ourselves, appropriately proprioceptive. Olson characterized proprioceptive knowing in terms of the dancer who envisions her own back as she moves in space, who he later imagined as a slug of type, the apparatus that knows its own form mirrored and reversed. At that time he could not imagine the electronic character that knows itself in each bit of arrayed, pixillated light form momentarily etched upon the phosphor of the VR goggle.

Nor could he or we imagine the disembodied embodiment of the bioapparatus, the out-of-body experience of ourselves as light. In the dancer's gravity of our constant fall into the terminal embrace of history, we hold ourselves as tellers of our own stories and fall into a weightlessness so surely there that we feel we could lay our heads against the scent of the past itself, nestled if not in memory then in recurrence. Uniquely ourselves, finally native, we remain foreign in this saving fall. "The paradox is that the foreigner wishes to be alone but with partners," says Kristeva, "and yet none is willing to join him in the space of his uniqueness" (1991). What befalls us is not the unbearable lightness of being but, rather, the being of unbearable lightness.

A Feel for Prose: Interstitial Links and the Contours of Hypertext

First Interstitial

Going back to, trying to read, Buffalo, as on a bus through Mayan Mexico, always first the family, *they wear their flesh with that difference that the understanding that it is common leads to,* and then Charles Olson, *what really matters: that a thing, any thing, impinges on us by a more important fact, its self-existence, without reference to any other thing,* reading perhaps, as on a bus, thinking about a feel for prose or why *reading ourselves . . . in a technological world awakens us to our roles, and our complicity, in the world,* Nancy Kaplan pressing against Freire.[1] *Our practical work must begin with reading the world, but it must not end there, acquiescing to that apparently authoritative text in front of us. Rather, [we] must actively appropriate the world text and thus reinscribe–re-vision–the technology of the word.* Even so, what can we touch; what do we think we touch in touch with prose? *The problem with getting inside the act of reading,* Jane Yellowlees Douglas, *is its ubiquity–there's no escaping it, and, like any environment that we are overly familiar with, we no longer see it. When we read print narratives we arrive already equipped with a full repertoire of reactions and strategies. . . . We never come face to face with the ground zero of reading.* The method here is Charles Bernstein's: *when trying to understand the nature of different media, it is often useful to think about what characterizes one medium in a way that distinguishes it from all other media–what is its essence, what can it do that no other medium can do?*

✪

(Contour)

We only sometimes recover a medieval memory of the touch of the book—recto and verso, verso and recto—and then for no more duration than a child's memory of the play of headlights through a window across the fern-fingered wallpaper in the dark moments after the bedtime story. Recall the book as play of light on light, page whispering as it settles on page. Since in the primal act of reading we touch the bare recto, the frontispiece, first, always on every recto thereafter is the sense of setting off from it, the verso awaiting as if in ambush, its print on the veiled page beneath onrushing and implicate, mirrored, suspicious of our simplicities. The constant immanence of the verso, taken together with our right-sweeping eyes, causes us when we begin to read leftside/rightside across the open folio of facing leaves to give ourselves less to the recto. We find ourselves thinly scanning it and then just as thinly circling back and guiltily (re)constructing argument or logic for its thicker paragraphs lest we miss some click of syntactic tumblers, some kiss of goddess or prince, that will open up what lies before us. Feeling rushed into the surf and unwilling to be either thrown over or shown up by the overleaf, we find ourselves at the page bottom almost childishly contrite. We launch across the abyss with a quick recital of the last line or two, a hold-your-nose paternoster attempt at memory, as children do in the chalky moments before a test, taking in one desperate fact before the sudden sigh and slap of the closing volume. We pray for rectitude. We pray that what we momentarily imprint upon the consciousness will hold in the whispery chant of memory and that memory, in turn, will transubstantiate it into a pale version of the now closed page. We pray to surface.

Yet once the turn from leaf to leaf is made the previous simplicities become ominously reversed. History appears in the form of already read, overthrown pages that appear beneath the newly upturned verso. A backward hegemony of runic inscription shows through the page in elven, dark shadows beneath the opacity of the fresh verso. This is the past embodied, patterned and printed upon our backs yet undecipherable, susceptible to retrieval yet never again new as the once turned recto.[2] We have always missed something there, some promise that, when revealed to us, now will pass before us unrecognized and unseized because of what we did not once see. The foreshadowing surface of the thigh, belly, cheek, and neck, the hollow of the back forever lost. We are always printing: each turn of the page opening another text unreasonably composed of the same gestures. We are always printing. No gesture ever the initial caress.

Despite what is buried, found, and lost beneath it, the verso as we turn to it always seems more light suffused, a casement, a lattice, not the striated but the smooth space, sweep of beach not edge of shore.[3] Across this broad strand of the verso the opposing recto reflects a buttery, reticulated light, one that glows from within the molten block of text up and outward through the recto's now smooth face, filtered only by the indistinct fingers and twigs, which give shape to the seemingly blurred certainties of its as yet undemanding, open text.[4] Thus, we set off jauntily on the verso with what some might call foolish confidence. Its surface offers us a crispness, illumination, articulation, the feel of vellum.

It is natural that the book shows itself to us as buried treasure, something to be dug through or up. The frontispiece is the insignia of the text as ingot, the block of it lumped beneath, fused by the heat of intellect, just as suddenly showing itself as flesh on flesh, *millefeuille,* gold leaf and labial, sideways symmetrically facing lips, gracefully flapping Dumbo ears taking flight like frigate birds over treasure ships sunk on the glittering shoals. Text becomes a paradoxical encrustation, a texture untouched, barnacle, mussel, echo; recto and verso addressing each other as the hither and thithering shores of the Liffey yet, despite the gutter they share, as far apart as elm from stone.

For the book always takes the form of its occasion for us. In prose at least its text holds no intentional relation to its space: the line we search for and that we have held in memory we know proprioceptively exists in a narrow sentence that sits like a grainy sliver of light lying just to the left part of an upper third of a right-side page of ominously bulky, slow, even cloudlike, paragraphs, which we recall by weight as embedded something over a third through the volume as it lays in the hand. We are always printing. Of the attributes that embody our imprinted memory of this sentence, the author can be imagined to have intended perhaps its relative depth in the volume—the poured strata within the ingot—as well as its play of alternation and lightness—the density local to the dark clouds and silvery sliver that float and etch themselves across the gridwork of the collected pages. All else is apparatus, cultural artifact or (as) economic product.

The narrow-spined volume clasped in the tight pocket of the airplane seatback and the edged weight and dusty texture of the novel against the stomach are incidental. Occasion takes form more intentionally. We inevitably construct the book for ourselves as a texture, varicose and hive-like, worming through the artifact, aswim in an intermittence that is for us enchained like fingers of kelp floating in an irregular pattern across the leftside-rightside waves of recto and verso. The linear, whether as the stability of hours or, ambiguously, the sequence we dream to overturn, is likewise an apparatus, a lie.[5] It is the fictional underlayment, the theatrical flat, the grid against which we project our imprint of the geometry of the text; and beneath which, fast as a sparrows heart, always beats this forgotten rhythm, recto/verso, of the enacted reading; and beyond which, as the play of light on light, the projective encompasses the object, wraps around it and—hiving—gives dimension, much as a movie slips off a screen or the eye from the curve of flesh: what the dancer knows, as Olson has it, herself seen through, her own back moving forward.

Second Interstitial

Halfway. Into the blue again,[6] coming down into Hamilton, the white Citroën pressed against the mountain like the curve of breast against the sleeping torso. In a haze of dark smoke and the yellow-green light of the cabin's electronic displays, the pomo duo, mad Bruno and Julia the therapist, smoke and argue through the night. On the radio someone is saying it doesn't do to argue for or against the possibility of an electronic or any text. Latour looks away from the mountain curve as, quietly, he seems to disagree. *Three possible readings all lead to the demise of the text. If you give up, the text does not count and might as well not have been written at all. If you go along, you believe it so much that it is quickly abstracted . . . and sinks into tacit practice. Lastly, if you work through the author's trials, you quit the text and enter the laboratory.* Tires barely hold against the centrifugal force of the mountainside. The radio persists: product or artifact, people do read from screens and write for them. The electronic book is left to enact either an orthodoxy or the laboratory. In this sense it *is* Latour's knowledge in action, or what, on the radio, Eco calls "work in movement," an auto barely in control. In the briefly dead air everyone smiles at the coincidence, and so when Kristeva speaks it is not clear whether she responds to Bruno or the radio. *An otherness barely touched upon and that already moves away,* she says. Perhaps she speaks to me as, letting the days go by, in silence I watch Hamilton unfold below us in scattered lights toward the dark lake. On the radio there is soft talk of toccatas and fugues. The method this time is Kristeva's: *Let us not seek to solidify, to turn the otherness... into a thing. Let us merely touch it, brush by it, without giving it a permanent structure. Simply sketching out its perpetual motion through some of its variegated aspects spread out before our eyes today, through some of its former changing representations scattered throughout history....* On the radio harpsichord; above and behind us, where we have come along the shelf of the road across the cliff face in the darkness, is the waterfall, flowing under ground; and on the dashboard before us, map or hieroglyph, is the electronic text.

✿

(Contour)

We as yet read the electronic text always as an irritation, an eye mote lapsing into a smeared band of reticulated light, a stereographic projection at a middle distance, focusing on it as on the glass through which the garden is seen.[7] Text on the screen is conspicuously nonprint, unheavy, undark, dry, unimprinted, prone to sailing off. Yet it is just as conspicuously not also the eye's distant white noise of television, neither the undifferentiated blur, the illusionary road shimmer, nor the embodied presence, not the angelic hum and play of light that television draws in air. Instead, with the electronic text there's undeniably a thing there on the screen. Like children's rolypoly pushover toys it somehow stays where it is when you scroll it, though it is always momentarily lost, or the eye is. It always threatens to go away. Even when static, electronic text seems frantic. It constantly reenacts its making, bobbing up from the dark beneath with the same buoyant insistence it has when you type it. Blurts of light emerge; more linear than ever print is in their constant inscription, the text always pushing the cursor forward or dancing above a block of light like singalong lyrics in old movie houses. In the midst of all this meaningless action there is always an air of monkish annoyance, the slow grace of something dull at work. We sense that each character is being drawn serif by serif, arduously and curvilinearly, each block of words insistently renewed, the play of light periodic, pixels Uzi etched boustrophedonic, how oxen walk in the glare of day.[8] Oh, we aren't at all happy with it, though it seems, like the circus, we ought to like it.

Print stays itself; electronic text replaces itself. If with the book we are always printing—always opening another text unreasonably composed of the same gestures—with electronic text we are always painting, each screen unreasonably washing away what was and replacing it with itself. The eye never rests upon it, though we are apt to feel the finger can touch it. The feel for electronic text is constant and plastic, the transubstantiated smear that, like Silly Putty, gives way to liquid or, like a painter's acrylics, forms into still encapsulated light. We are always painting. The electronic is not at all the touch of the uncertain reader, who, like a child poking at a line of ants or lining up raisins runs a finger along each cast line of print. Rather, it is certain touch, like holding moths, feeling the velvety resistance as the wing's scales slip off against the press of the fingers, dusting the whorls of the fingertip at whose root the hooked tangle of a zipper row of pincers clings, below that next to the clasped, now transparent wings, dangles the segmented leathery body, the husk of the idea dangling as palpable as Plato's dry forms, yet a worm still wet within, exploding with damp, phosphor light should you squish it open.

Not words but wings. Electronic text replaces itself with something of the pendulous, anxious instant that follows when the lecturer calls "Next slide, please" to the unseen projectionist. Yet screen text lacks both the mechanical and the optical determinacy with which the gelled image is summoned and illuminated—caught in a click, gripped, the worm gear slipping it first past then into focus, light rays bent like reinforcing rods awkwardly into place along the principal axis, where the image is formed from a gurry of concrete light. On the printed page we have the illusion that the words are out of focus and that they shape themselves as we narrow in on them; electronic text opens fully focused and impatient; the words await us. The speed of the changing screen stays with us as we read, and our eyes water and burn in the glare of such clarity. Despite the wing's buckle of swift focus as text opens on screen, it doesn't etch itself upon us. More photographic than typographic, electronic text emerges as the image does from the chemical bath in the darkroom, in present tense, fully focused before its birth, ablink, the whiteness not defining but interwoven. Thus, though it dismays filmmakers, when the syntagma of the visual dissolve is misappropriated into a textual transitional device for the computer screen, it is nonetheless fitting. To the filmmaker the visual dissolve signifies the ellipsis of the historical present, the shift of tense to time passing. Electronic text is, instead, the constantly replaced present tense, the interwovenness, the interstitial, which the dissolve, rather than signifying, enacts.

Electronic text appears as dissipate mist; the swift dissolve of fog burning from canyon rim is the visual quality of this shift. Print stays itself; electronic text replaces itself. What poets think to hate about the computer screen is the loss of an artifact, not the temporality of words but, instead, the artifact of words' temporality. Our memory of text on screen is as intricate as light or bees or the gold gravure of circuit on silicon, yet it always renews itself in a context less contiguous than lantern slides, less stratified than the memory of print. Under it, as under the tongue is the throated darkness, always the dark of ancient water cuts through the ephemeral lightness of stone. Etched, hive-like, varicose the canyon of the world serpent, *ouriboros*, is the dark, defining abyss, a desert of silicon across which the constantly replaced present strobes like lightning. At canyon bottom the silted electrons of sand are possessed of a memory predating the one they have been formed to serve. Atom recalling granule, granule stone, stone the great mountain, mountain the first home.

Third Interstitial

At the Rainbow Bridge we are met by the hypertext publisher (not the poet) Bernstein, the next generation's James Laughlin, who has exchanged the olive drab flak jacket and black paratrooper's boots that he travels in for the black suit he wears to academic conferences, business meetings, and, one supposes, the seasonal occasions of family. Boyishly parted straight hair, piercing eyes, and an aquiline nose make him look exactly like the photos of the young man Picasso. A bad ear, a chemist's impatience with the volatile, and the vested black suit give him a Mandarin's manners and bearing. *It's explaining the revolution to the bystanders that befuddles me.* He gets into the car next to me, folding like a crane into dark water. Bruno and Julia smile to him. The method here is his. *I'm still casting around for the bridge rhetoric, the little bits about "discarding bourgeois sentimentality," "plein air," "painting light." So that when someone is walking around the gallery and says, "Why doesn't young Monet bother to finish anything?" I have something comforting and useful to say.* Once again smuggling Irish in across the border from Canada, their insistent foreignness will conceal me.

✿

(Contour)

We are interested in a text as a place of encounter in which we continually create the future. Yet we desire our future as we experience it to remain the mystery it is as we anticipate it.[9] Our encounter with the future text thus carries with it what might be called the melancholy of history. The awkward light of dawn isolates into angles and instances a half-illuminated world of the shapes we desire to project upon its shadows. We are always printing; we are always painting. Before long the constantly replaced present will dissolve the seamlessly imprinted future within it. Insofar as we can glimpse them both at the moment of dissolution, we experience the momentary advantage of the awkwardness of change. We are allowed to see what Latour calls the "new object" in the moment before it enters the seamless cycle of resumption and progression of the historical day. "Narrative," Susan Howe writes, "expanding, contracting, dissolving. Nearer to know less before afterward schism in sum. No hierarchy, no notion of polarity. Perception of an object means loosing and losing it" (1985, 23).

Confronted by the new medium of multiple (electronic) texts, we consider that the way we read imaginative writing now threatens to become the sometimes tiresome process that in our desire we once claimed it to be. Multiple fictions quite literally require collaboration by the reader of the hypertext to give meaning to the texts through her constant textual intervention and shaping, her construction of successive interpretive frameworks, and her responses as a reader. This is to say that with electronic texts slippery notions, which have become hallmarks of postmodern criticism—like the notions of interpretive community or intertextuality—no longer enjoy the luxury and repose of theory.

E.g., while multiple and print fictions are alike in that the "pages" do not turn without the reader's intervention, they are radically different in what follows when the reader turns them. Not only do the choices a reader makes in an electronic text govern what she next sees, but they also unfold patterns for her discovery of the narrative in much the same way that conversational cues shape our discovery of one another. While we may argue that an unfolding pattern of discovery of the text is also present in print, hypertext both embodies and is itself solely embodied by what in print is an invisible process. The screen enacts the ground zero of reading. There the reader of a hypertext not only chooses the order of what she reads, but her choices, in fact, become what it is. The text continually rewrites itself and becomes what I term the constructive hypertext: a version of what it is becoming, a structure for what does not yet exist. Yet even this version, this structure, is continually being replaced, either by itself or by what it has become—either or both of them thrust into present tense by the relentless cycle of painted light.

Since we as yet think ourselves unable to read or write in the substance of light itself, the locus for this inscription resides in the interstices within any word. In this sense the electronic word recapitulates as light in space the radical linearity of Ong's "sounded word as power and action" (1982, 32).[10] It is as apt to evolve before it forms, as apt to dissolve before it finishes. It takes our constant and attentive interaction to maintain even the simulacrum of static text. The future, too, requires as much of us and has for some time.

Electronic texts present themselves in the medium of their dissolution. They are read where they are written; they are written as they are read. In this way they evoke and mimic manuscript culture. In another way electronic texts mimic oral culture in their aggregative rather than analytical qualities.[11] Unlike both, but like light, they are colored by both their own complementary and additive qualities as well as by the subtractive interventions of what they reflect. We may call this quality the network culture or polylogue. For the writer in the daily polylogue of the network culture, texts are filtered from a constantly emitted spectrum of electronic quanta including e-mail messages, letters, other texts, network postings, stored electronic manuscripts, and writing newly generated. The same tools that (like Storyspace) are the medium of written hypertext also enable, enact, and map the polylogue for the writer. Network culture thus situates the intertextual confluence of writing-conscious-of-itself. Simply because it is easy to do so, writers are more likely to be conscious of the polylogous texts that their own texts impinge on, border, float upon, or dissolve within them.

Network culture likewise situates the interpretive community. Multiple fictions have no print equivalent for the experience of how their sequence is generated, and there is not even a mathematical possibility of printing their variations. The pages do not turn alike for any reader. Since the only condition in which a reader reads the same sequence of text as someone else is if she consults with him, even the slippery notion of interpretive community is embodied. Meanings sound in the music of these contrasting characterizations.[12] An author "performing" a multiple fiction in a public reading (in which the sequence of texts is governed by one audience member who guides and/or responds to the choices other audience members make) likewise depends upon readers to form the text that evolves and is read.

Rendered in the pure light of ground zero, the intertextual polylogue, too, is perceived as the sometimes tiresome process that our desire claimed it to be. Yet even though the new object seems to be replacement itself, it is neither simulation nor seduction, since with electronic text there's undeniably a rolypoly pushover thing there on the screen. "There is a great difference between the infinite and the inaccessible," Jay Bolter writes: "Electronic writing with its graphical representations of structure encourages us to think that intertextual relations can indeed be mapped out" (1991, 202). Once mapped we have to go: to think is to voyage. This is the nature of the new medium, the new object. Print stays itself; electronic text replaces itself. "Voyage in place: that is the name of all intensities," say Deleuze and Guattari, "even if they develop also in extension": "To think is to voyage . . . what distinguishes the two kinds of voyages is . . . the mode of spatialization, the manner of being in space, of being for space" (1987, 482).

Electronic text is the mode of being for space within the network culture. What frightens us about being for space is the loss of the analogue, the ascent of the digital. A student writes in his journal (it has been composed on a computer and printed in three-column format in the guise of a newspaper): "With a record I could get some sound off it even if I had to use a bamboo needle and a klaxon speaker like we did once in Nam. With a CD there's nothing I can do to get at those zeros and ones." He seems unaware of the ironies in this parable of war against the unseen, especially now that the war machine attempts to appropriate the electronic text as the "smart bomb."

The zeros and ones of the digital mean by merely being, and so the digital is an instance of Latour's new object. We fear yet another of the claims of our desire, that reality itself will be altered in a way we cannot get at. No matter that the wave forms of analog sound were once as much a new object, once as green a thing as the new bamboo shoot. "Laboratories," says Latour, "are now powerful enough to define reality.... [R]eality as the latin word *res* indicates, is what *resists*. What does it resist? *Trials of strength*. If, in a given situation, no dissenter is able to modify the shape of the new object, then that's it, it *is* reality" (1989, 93).

Electronic text—the topographic, truly digital writing[13]—even now resists attempts to wrestle it back into analogue or modify its shape into the shape of print. Its resistance is its malleability. We cannot get at it because, being for space not in space, it will not stand still. As Richard Lanham puts it, "Digitization both desubstantiates a work of art and subjects it to perpetual immanent metamorphosis from one sense-dimension to another" (1989, 268). Not the unbearable lightness of being but, rather, the being of unbearable lightness.

Here is a nest of paradoxes. If we wish to know the text our reader has written through her choices, we will have to reciprocate. She can only know by our choices what distinguishes her own. We construct the electronic text by our choices, but we only come to know what we have written by understanding the choices of others. To get at the zeros and ones that are the music of our spheres is to enter the network. The network culture is, like Latour's scientific writing, *more* social than so-called normal social ties. We understand from the third person what we have written in the first person but only in the process of reading the second person. The electronic culture estranges us not from the familiar but from the privilege of isolate individuality. To think is to voyage; to voyage is to willingly become foreign. "The paradox is that the foreigner wishes to be alone but with partners," says Kristeva, "and yet none is willing to join him in the space of his uniqueness" (1991, 12).

Last Interstitial

Home now. Days later. Home? Now? Buffalo. A foreigner alone. Everyone gone. Julia. Boris. Bernstein. Mother, father dead. At the front of the room an authority embodied. One last section. Only pages away from the apparent closure, the audience perhaps tiring, else champing with the sour metallic taste of rising questions. Who can say now, when I write this, whether the future this text claims will take the form of its desire? Authority as always some sort of cyborg. Our memories externalized in an apparatus: snapshots or texts painted in plein air by Teleprompters, the zeros and ones of the digital. Everything he says now might as well be narrated in the three-second delay of broadcast programs. The printed text before him could as well come from the hidden electronic text on the mini-CD Discman and fed to his ear by the infrared wireless earphones Sony will market late this year; everything fed to his synapses via a thin surgical platinum wire antenna embedded behind his ear receiving the Data-Personal Communication Services signals Apple proposes to beam worldwide soon. We have lived to the time of the village idiot, when we wisely heed the messages received from the fillings of our teeth. The method here—paragnosis (Stuart Moulthrop's term). The voice on the wireless beam his? Whose? *Deconstructive hypertext would apply itself to an existing linear structure in order to nullify its apparent closure. . . . [It] opens the prior text to further reassemblies or acts of linkage: it becomes a structure for analyses that do not yet exist. [. . .] [A]pplied to hypertexts themselves . . . deconstructive hypertext might remind us that any system, even (or especially) one that advertises its own contingency, can have its authority called into question. It might demonstrate that every post-Gutenberg text includes a dimension of absence or indefiniteness—an element "that does not yet exist"—which is its opening to further discourse.*

✪

(Contour)

In their multiple fiction, *Izme Pass,* Caroline Guyer and Martha Petry consciously map and mediate what they describe as a "deconstruction of priority... the purely arbitrary construction we decided to work within... a triad of existing texts... among which we began to weave a fourth, new work." As a genuine product of network culture, *Izme Pass* enacts "how things are connected, not connection as conceptual negative space, but connection itself being a figure against the ground of writing" (1991b, 82). They enact a text that anticipates Moulthrop's deconstructive hypertext, since it is itself a present-tense reading of other texts.

The contingency and indefiniteness of hypertext invites this kind of reconstruction and reconsideration of figure and ground in narratological terms. Guyer and Petry's sense of "connection itself being a figure" offers a key to what I want to call a poetic of contours. Contour is ground as the mode of spatialization, the manner of being both in space and for space. A contour is the space of inscription for a reader, the emerging surface of the constructive text as it is shaped by its reading. "A turn *toward* a destination," Terence Harpold writes in building upon George Landow's "rhetoric of arrivals and departures" in hypertext, "is also a turn *away from* an origin; turning encircles a place that is neither origin nor destination, and the shape of the turn divides you from its center": "The different names for that turn describe... the contour of a place you never got to" (1990, 3). Coextensivity and depth, taken together, trace the constantly replaced present, the place you never got to, which underlies a reader's (and writer's) experience of hypertext. There at the contour the reader's inscription enacts her perception of coextensivity and of depth.

We might characterize coextensivity as the replacement of one writing by another. Hypertexts are read where they are written and written as they are read. We recognize coextensivity in our ability to reach any other contour from some point of a hypertext. I.e., coextensivity is the degree of impingement and dissolution among elements of a hypertext. Coextensivity is the manner of being for space. Our eyes rewrite each surface of a sculptural or any other smooth three-dimensional morphological form, replacing surfaces of coextensive contours. Depth, then, can be characterized as the contour-to-contour inscriptions or links that precede, follow, or reside at any interstice along the current contour of a hypertext. Depth is the manner of being in space, the capacity for replacement among elements of a hypertext. Our eyes read depth as the dimension of absence or indefiniteness that opens to further discourse. It is this same striated dimension that is seen in relief maps and other two-dimensional morphological forms of desire.

The ability to perceive both coextensivity and depth instantiates the interactivity of electronic texts, making the "structural strategy" of Umberto Eco's open text available for inscription. For Eco "the possibilities which the work's openness makes available always work within a given field of relations... the *work in movement* is the possibility of numerous personal interventions, but it is not an amorphous invitation to indiscriminate participation": "The invitation offers the performer the chance of oriented insertion into something which always remains the world intended by the author" (1989, 19). Yet the introduction of connection itself as a figure of the text dissolves imprinted future into replaced present, author into performer. The world intended by the author is a place of encounter in which we continually create the future, the interstitial link wherein we enact the replacement of one writing by another.

Our perception of the interstitial link seems to require some contour to perform as a base line that we measure ourselves against. Contour is at least initially perceived in voyage, in the occurrence of a text that disjuncts us from our oriented insertion into an apparent continuity. Deleuze and Guattari describe the binary series of the desiring machine as "linear in every direction... every 'object' presupposes the continuity of a flow, every flow, the fragmentation of the object" (1987, 399). In a simple discursive (what I call exploratory) hypertext,[14] someone reading an article about earthquakes might select the name Richter at a point in the text that begins to describe the San Andreas fault. The reader would then see a brief text, including a biography of Richter, immediately followed by another text with an explanation of the Richter scale. The biography and explanation thus appear without any indication in the original text that they would "normally" follow. Under these circumstances it is likely that the reader would perceive the biography as disjunct from the earthquake article and perhaps, likewise, perceive the explanation as disjunct from the biography. Yet the biography, because it comes first and in response to the selection of Richter, might seem less disjunct than the explanation, since the latter would appear without any active selection upon the reader's part. Finally, the perception of contour would be heightened if a further discussion of the San Andreas fault replaces the biography and explanation. Obviously, however, no such discussion is necessary for the reader to perceive the text as coextensive. We may argue, as Jane Yellowlees Douglas argues for multiple fictions, that coextensivity is present from the first:

> Most hypertext narratives have no single beginning. In Stuart Moulthrop's *Victory Garden*, readers are confronted with no fewer than three lists which seem, at first glance, to represent a sort of Table of Contents: "Places to Be," "Paths to Explore," and "Paths to Deplore." Unlike a Table of Contents, however, these lists do not represent a hierarchical map of the narrative, providing readers with a preview of the topics they will explore during their reading and the order in which they will experience them.... The first place or path in the list has no priority over any of the others—readers will not necessarily encounter it first, and need not encounter it at all, in the course of their reading. Each of the words or phrases, instead, acts as a contact point for readers entering the narrative. By choosing an intriguing word or particularly interesting phrase, readers find themselves launched on one of the many paths through the text. In print narratives, reading the Table of Contents—if there is one—is generally irrelevant to our experience of the narrative itself: our reading experience begins with the first words of the narrative and is completed by the last words on the last page. In *Victory Garden*, however, readers are unable to begin reading without browsing through the lists of places and paths and then selecting one. In this particular instance, actual, physical interaction takes place right from the outset of the reading experience. (Douglas 1994, in press)

Even so, the reader always begins somewhere, whether in the first or a subsequent reading of a hypertext. In the discursive example the contour would be "an article about earthquakes." Given her experience with print, the reader perceives that she can continue along this ground freely in a default, "automatic" sequential reading, and many hypertext systems include this ability as part of their design. In a hypertext system, however, the reader may also choose a link from the text of the initial ground, i.e., by selecting the word *Richter* in the example above. Theoretically, at least, the reader can choose links at any point in any text within a hypertext. So, when our reader does choose a link we may suppose that she either perceives a shift to another contour or perceives the figure of connection itself that orients her in relation to that ground. It is her gathering perception of depth and coextensivity that, in Moulthrop's words, "opens the prior text to further reassemblies or acts of linkage" (1991).

Whether we seek a poetic of contours or a deconstructive rhetoric, we must actively locate them at the interstices along the continuously replaced contour. Locating here must be understood in the double sense of interaction and enaction. I.e., we locate by inscription, forcing (or enforcing) the coextensivity of the text. Given the resistence of a system that advertises its own contingency (where Latour's laboratory is given over to beakers of self-organizing systems), we are forced to ask how the reader could possibly resist, appropriate, possess, and replace the figure of connection. "The systems that control the formats that determine the genres of our everyday life are inaccessible to us," writes Charles Bernstein: "We are structurally excluded from having access to the command structure: very few know the language, and even fewer can (re)write it" (1991).

The problem we face is how to write in the interstices. If the reader is programming, the reader is programmed. The program that underlies electronic text is print, the imprinted future, the ever recurring, never replaced, historical present. Unlike the system of the videogame, however, for which Charles Bernstein observes that, "even if we could rewrite these deep structures, the systems are hardwired in such a way as to prevent such tampering," the system of the electronic text cannot be successfully hardwired. Because it is a system based on recursiveness, the continual replacement of itself, it always must leave what the programmers call a back door, where it accomplishes its own rewriting, the interstitial link wherein we enact the replacement of one writing by another. The back door can be disguised in the seduction of technology (the hypermedia of Quicktime, expanded book, or virtual reality), but these productions only mean to distract us by raising the cost of entry, the cover charge, for the floor show. They only affect those who wish to see. Thus, the viciousness of the attack on writing that inevitably accompanies these posttechnological productions ("Why read when you can see? why see when you can experience?" says Jaron Lanier, the flyboy saint of virtual reality) is meant to lure us away from the unlatched access bay to the self-organizing system.

The electronic text presently lies at the equipoise of colonization or voyage. As a colonization, it can easily become a programmed empowerment, a magic capture, or what Richard Grusin has recently called the trivial pursuit of the matrix. Moulthrop warns that hypertext allows a state apparatus to "subject itself to trivial critiques in order to pre-empt any real questioning of authority" (21). Electronic text might thus become, one supposes, disembodied internalizations of the nightly floor show of "Nightline" or "Firing Line." Even more threatening than the text as spectacle, however, is the seduction of isolate individuality that the screen enacts as it hypnotically washes away what was and replaces it with itself. "Solitude is perhaps the only word that has no meaning," writes Kristeva: "Without other, without guidepost, it cannot bear the difference that, alone, discriminates and makes sense" (12).

Network culture is more social than so-called normal social ties, yet, mesmerized by the successive contours of the text, the reader can find herself engaged in a process of annotation and addition, rather than one of replacement. In this case each addition she makes falls as a terminal node on an underlying graph, a note to an unseen, underlying text. Though empowerment veils itself in the intransitive, something purports to instruct this occasion of power. An imprinted future not only awaits but also anticipates and appropriates her additions. The reader is the audience of programming in the double sense of instructed machine behavior and the thing that, like situation comedy or docudrama, is shown on the screen. In this laboratory the dissenter's tests have returned false-positive: it turns out that she is unable to modify the shape of the new object, and so the real persists. Her additions are marginal, not coextensive. Interaction does not enact the replacement of one writing by another but, rather, conserves authority. We play ping to the pong of an unseen program.

A fully coextensive, truly constructive electronic text will present the reader with opportunities for capturing the figure of connection at its interstices. The evolving contour must be manifest for the reader so she can recognize, resist, appropriate, possess, replace, and deploy the existing contour not just in its logic and nuance but also in its plasticity. She should be able to mold and extend the existing structure at each point of replacement and to transform it to her own uses. As she gets a feel for the contours of the text, she should be able to predict how its existing elements conform to the contour of her replacements. While not merely taking on but surrendering its contour to the reader's inscription, the transformed and replaced contour should nonetheless conserve its depth, continuing to hold the shape it had for previous readers beneath the finger that traces the replaced contours of the transformations made by successive readers. Rather than playing ping to the unseen, we shape the visible object of our longing.

Occasion takes form intentionally. Since we inevitably construct the book, too, as a texture, modes of spatialization obviously suggest themselves to readers of linear texts. A glossary, e.g., is a contour that might be said to exist below another contour; a table of contents might be said to exist above and an index below another contour; a commentary might be said to exist in parallel. Less intuitive contours are nonetheless fairly common to poetics: motifs may be said to exist as interweavings, or tropes might be said to exist as quilted interdependencies. While intuitive morphological relationships among contours are interesting and potentially exploitable in deriving a transitional rhetoric or poetic as a bridge from linear texts, contours enjoy the benefit of not being tied to traditional poetics or rhetorical forms. A contour provides a way to account for how a text routinely and radically shifts according to the reader's perception of its continuous replacement. These perceptual reversals—to borrow a term from Arnheim—are fundamental to the reader's perception of coextensivity and depth.

Contours alternately take the foreground and then surrender it like water. They become the background continuity, the current and flow, as the reader perceives a gathering sense of multiple uses of the hypertext and their interrelationship. In touch with the contours themselves, moving through them as a hand through the stream, channeling the flow like light through a prism, the coextensivity of the text becomes the weave and reversal of water. The reader also learns to recognize the perceptual reversals, the ripples in which one contour impinges upon another, taking and surrendering the perceived continuity. In so doing, she assumes control of both the alternation of contours and the multiply potentiated uses of the text. Capturing the flow in this way, channeling it, the reader turns the text to distinctive uses of her own, which she can float upon or navigate through. She begins to voyage, both in space and for space.

"I anticipate that our questions and answers will change in the asking," Martha Petry writes in the printed polylogue that accompanies *Izme Pass*, the weave of e-mail messages between its reader-authors:

> that we will have different utterances and tones depending on our emotional currents. I think of the writing as essentially flat, words on a page, icons on a screen, that become multidimensional as we layer, link, place, guard—our collaboration becoming a stream that cuts through the bedrock, a stream that is a continuously, synchronously, changing event where diving and stoneskipping, leaping across and contemplating, can happen simultaneously. (1991, 83)

We must listen so carefully to one another, how the stones splash.

Appendix: Descriptive Summaries to Chapter 4

In the following descriptive summaries thematic spaces appear in the following form, {*beginnings (8)*}, in which my title for the thematic space, "beginnings," appears first, followed by the number of interchanges in the section (8). In these summaries some sections are run together into long paragraphs both to retain the flavor of the interchange flow and to approximate embedded levels. The space, {*choices again (31)*}, in the HT Writers' summary is boldfaced to indicate that it is the subject of further linkplot analysis in chapter 4.

> Descriptive Summary—HT Writers. {*Agenda-setting (9)*}. Housekeeping and hellos. The group affirms its customs, makes certain that a particular discourse won't be constrained. {*Definite form (7)*} Bill attempts to set the topic to discussion of form, but the flow of introductions, delayed agenda-setting, etc., deflects him. He restates and slightly alters his theme. {*Resetting agenda (8)*} A short exchange asserts the accustomed patterns, i.e., "say[ing] things that are totally irrelevant," against Bill's foray, which he repeats. {*Definite form2 (7)*} Dawnkiller picks up on Bill's theme of form and offers a definition of hypertext, "emotion . . . seems to be the embodiment of hypertext." Zøʃó attempts to pick up my strand about visual form, making a distinction between hypertext reading and writing. The others parody these discussions with a continued thread about flies (and pies?) in which Boomer connects the perception/interruption of a fly buzz with hypertext reading as Dawnkiller raises the issue of real vs. virtual books. {*The "book" (8)*} Moon picks up on a note of Dawnkiller's referring to a past Interchange session regarding hypertext and writing, the transcript of which had been lost. (Moon's report of that previous exchange in some sense precipitated this session.) Now

she proposes to raise the issue of books vs. hypertext again. Others, however, seem more inclined to follow on either Bill's discussion of form or my question about visual writing. Dawnkiller merges the discussions in a brilliant gambit regarding internal form and the nature of the book. Meanwhile, Bill's and ZøʃÓ's talk about beginnings of hypertexts will both provide the next theme and merge discussion. *{beginnings (8)}* The talk of beginnings and endings (as I discover only after analyzing them both in sequence) is interwoven with a consideration of how and whether we "let" the reader have choices. Dawnkiller's meditation on book and hypertext temporarily ends that thread. The group sets off to discuss beginnings and their relationship to readers' choices, or "directions." The discussion is ambiguated by Dawnkiller's connection of life and text. Beginnings inevitably lead to discussion of endings, though Bill formally sets the topic as "closure." *{endings (11)}* The discussion of endings is interwoven with a consideration of "letting" the reader have choices, which has also interwoven the discussion of beginnings. The agenda setting was probably mine, but the group passes the topic along through their discussions, having the notions of beginning and ending inform their emerging understanding of reader choices. *{future of books (21)}* A discussion of the future of books (ranging from what forms, i.e., CD or tape, to why libraries burn them) continues to be interwoven with one about choices, which I attempt to agenda-set into a discussion of hypertext methodology (how we disclose choices to readers). Only when Stephanie, the newest hypertext writer in the group, makes a distinction between hypertext and the writing she learned in high school does the discussion turn. I realize these last three sections of the analysis come under the broad umbrella/canopies/arcs of the future of the book that Moon set early on. Their multiple structure calls to mind Martha Petry's admonishment that, in hypertext collaboration, "we must listen so carefully to each other, how the stones splash" (Guyer and Petry 1991b, 83).

{choices again (31)} The weave continues (crescendoes) with a four-level discussion of (1) how to disclose choices to a reader and whether the reader can actually take control, (2) what will be the future of the book, (3) why did the college library burn books removed from the collection, (4) what is different about reading and writing hypertext? As a result (or so I hypothesize), I collect more spaces into the section of the analysis. Perhaps, however, it is only that I am already tiring with the more granular analysis, only a third of the way through the transcript. In any case Stephanie finds a way to enter the discussion in this section (and each time is mentored by Bill). Moon is seemingly (naturally?) unaware that the discussion continues under the "whither books" umbrella she has raised much earlier, and so she restates it formally here midway, again referring to the lost, past Interchange session. Wabbit and Dawntreader tie teaching, hypertext, and the future of the book in a thread about knowledge, partly in response to Stephanie. {*reader writing (26)*} The discussion of letting/leaving choices for the reader matures and mingles with the strand regarding the future of the book, including the various ways books are read and the kinds of access they allow. Wabbit takes on the role of the discussion leader almost transparently, cuing people's discussion. Z's agenda-setting question about whether the readers' choices change writers' creation is discussed from several angles, including Dawnkiller's characterization of the book as a "sacred . . . containment of knowledge" and Wabbit's sense that "a linear book is experienced individually and differently" and thus may be hypertext. Uther advances a notion of things that the writer may not want the reader to change (a notion Bill has raised much before in his idea of "theme"). By and large the discussion here is wide ranging and engaged, the group largely possessed of the question at hand. {*hypertext and life (29)*} The discussion blossoms into parallels between hypertext and life as well as alternate (preexisting) forms of hypertext such as Post-it notes on walls, etc. Bill launches this strand with a heartfelt meditation on new forms and the fear of change (his paired comment followed by a deprecatory, apologetic comment, a Discourse pattern). Dawnkiller ties real choices in hypertext and real life together in the notion of "true interaction" (recalling my selfish interaction [Joyce 1991a] or Andy Lippman's characterization of it as "mutual and simultaneous activity on the part of two participants, usually working toward some goal, but not necessarily" [reported in Brand 1987, 46]). {*other forms (26)*} The discussion of new and alternative forms (Post-its or paper texts linked with actual threads through the paper) and Bill's sense of mind & art are taken up in an almost dreamlike, poetically linked meditation on art, technology, reality, illusion, and mind. Even Z's joking request for the group's help in deciding what medication to take is eventually linked by moon to hypertext. So too, when Boomer, Uther, and others consider the missing eye contact in technology (something perhaps a Post-it wall would retain?), that thread is somehow linked or gathered into a consideration of the hormonal level of machines. The section is nearly inexplicable, almost certainly a real meditation of the themes so far. In the midst of it I have to sign off. {*cat's away (30)*} With my exit the group settles into a series of exchanges that largely concern what Boomer calls "interchange therapy." They are used to one another in Interchange and used to one another in community. There is some tying up loose ends about computer hormones, joking about music, and a number of attempts at policing/focusing/maintaining the discourse, ranging from moon's delightful slip "no multiple maning," which she quickly amends to "meaning" for this largely male group. In a two-way conversation between Bill and Steph regarding her boyfriend the others play the part of chorus (viz., Dawnkiller's "Lost in Steph."... "(Her mind)" following Z's agenda setting: "Where went the hypertext?" Bill tries to link the two strands in paired comments: "He needs to give himself a few key points to help himself. no one can give a person everything" and "no story gives you evrything no hypertext speaks all the words" On first analysis I think this linkage unsuccessful and attribute the next section's turn to consideration of hypertext to Lestat's entry, a skeptic about hypertext who is a tutor but not also in the hypertext writers' group assembled here. Yet I notice that, just after Bill leaves, moon, Boomer, and Dawnkiller quite consciously refocus on hypertext well before Lestat enters. A mock-formal exchange between moon and Boomer affirms the group and topic: "Boomer Petway: Hi moon//moon: hypertext maybe gives you more of a chance than a regular story.//Boomer Petway: tell me more moon//Dawnkiller: Maybe.//Boomer Petway: I hate when interchange is dominated by personal problems when a formal topic is presented." {*reforming discussion (25)*} A number of others have signed off. Lestat joins the conference (logging in from another room out of sight of the group), and a conscious sequence of interchanges focuses on agenda setting and summarizing. Lestat apparently knows that this session is in some way connected to the lost session with its extended discussion (echoed throughout here) of the library book burning as symbolic of the devaluation of books with the onset of technologies. Lestat interestingly sees himself as moon's "adversary," although she has been the rabid defender of books for readers and hypertext for writers. As Lestat engages in a twenty questions attempt to pick up the strand of this now hour-long discussion, moon suddenly links the idea of communal "therapy" with her perception of hypertext as a necessary form for her as a writer, affirming the community and this discourse in the process:

> okay, there is a person in here whom I recently wrote some "stuff" about. a friend, I suppose... in a way. I kind of watched his actions, for months, in reaction to losing someone he really loved, then thought of how to write about it. it was hard. I couldn't just sit and write continuos thoughts. they flew all around. thats where hypertext really helped.

{*hypertext critic (47)*} The consideration of Lestat's "opposition" takes something of the form of a Platonic dialogue, sharpening consideration of its subject and, in the process, shifting its meaning subtly and unalterably. Lestat's resistance to reading on screen gives way to his argument that hypertexts lack quality in their writing, especially as measured against "real books." Moon and Boomer question whether Lestat has ever written in hypertext, and he admits to being somewhat curious, especially as it functions in an "annotative" way or like a "text video game." They wonder together what changes in reading technology will come in the future and whether hypertext is a "fad." Moon assumes a role of defending hypertext as a writer's medium but not as a reader's (both because it is not portable and because she does not want to see the book replaced). This puts Lestat (who it seems in the previous spring's discourse took the "adversarial" role that book burning is okay because there are too many books) in the position of defending hypertexts as less wasteful delivery vehicles for technical writing and education. Moon agrees about tech writing but still is loath to see the book go. An interwoven strand throughout this section reconsiders the question of the writer's intentions. Early on Lestat argues that "an author has a certain direction s/he wants to move the reader in." Uther and Lestat thereafter conduct an interwoven dialogue about reading order and how the reader forms the text, which culminates in what the fish lady said. {*too many books (17)*} A brief, quirky, poetic section considers how best to maintain information for the ages, whether as book, in braille for a posttechnological age of darkness, as disk archive, or even through a revival of bardic tradition. Uther unconsciously defines the bardic in much the same way he defended hypertext to Lestat. "Bards are still around," Uther says when Lestat claims books made them obsolete: "They add qualities to the stories they sing that aren't in plain black and white." In the previous section he has said that hypertext "is just a way of deepening the feelings you can get out of a story." The section ends on an ironic note as Lestat muses on how the new technology of the screen removes him from interaction with the others around him in the class he is either taking or consulting in while he participates from afar in this interchange. {*reading screens (25)*} Following Lestat's comment about how the screen, unlike the book, absorbs him in an antisocial way, two strands are interwoven by turns. In one Lestat and moon discuss the unique nature of electronic text as well as kinds of absorbing reading experiences in print. Meanwhile, Lestat and Uther discuss access to, requirements of, and intimidation of technology. Lestat introduces a media-based critique of electronic text (the MTV-like short sentences, etc.), which Uther attempts to counter, first with a rhetorical suggestion, "It's hard to type a long sentence here and keep the attention of the person," and then with a hoary writing principle: "Ideas should be short and to the point." Lestat counters, "Tell that to Tolstoy!" Moon rehearses the groups earlier critique of the nonlinear quality of hypertext, but Lestat resists being so characterized, pointing out that he reads poems nonlinearly but stories in order. Uther ends the section by suggesting that "most stories are written to take you from point to point": "If you randomly move around you loose sense of what the story is."

{*reading HT's (12)*} A short, richly directed sequence on reading hypertexts follows from Uther's suggestion that "ytou loose sense of what the story is" in hypertext, a happy typographical pun that for me recalls Susan Howe's characterization of narrative, "expanding, contracting, dissolving": "No hierarchy, no notion of polarity. Perception of an object means loosing and losing it." In the section the group advances their understanding incrementally. Uther reformulates his previous idea about "point to point" stories into either a new understanding or (more likely) a special case definition for hypertexts, which "don't necessarily take you from point to point": "You take yourself where it is you want to go." The keynote for this inquiry is sounded by fisherwife woman (the same person as Fish Lady? a discourse name adlibbed and then not recalled exactly here?), who recalls her earlier image of hypertext in language that echoes William Carlos Williams's characterization of the poet in *Patterson*: "Hypertexts let you dig around in garbage, chew on bones, gnaw and howl at moons..." Echoing Uther's earlier "Htext is just a way of deepening the feelings you can get out of a story" is moon's view that hypertext "adds alot more emotion." {*writing HT's (25)*} Talk turns from reading hypertexts to writing them with a reprise of the earlier discussion of the larger group about "letting" readers have choices. Lestat wants to argue for an information model of the text, in which the reader is supposed to gather "points" the writer wants her to get. Moon, who sees the hypertext as a writer's medium, agrees, though Uther argues for a compromise view, something like the exploratory hypertext in which the reader has choices but reads "the whole thing" to get the meaning. When Lestat disagrees with this notion, Uther advances a view of what "good... careful" writing is in hypertext. Moon's question, "What is the difference between hypertext and regular writing?" at first strikes me as a kind of empty ameliorative turn-taking. It seems, like Uther's compromise position, an attempt at homogeneity, reducing the hypertext vs. book arguments to rules for good writing. However, moon, despite or because of her wariness about hypertext, is aware there are real differences. Thus, when Uther tries to argue that hypertext lets you "bounce around" and choose where you are going, moon doesn't let him off the hook, insisting that, if writing means to convey a point, there is no real difference. Uther counters that hypertext lets you skip parts of stories you don't like, but moon is tenacious: if writing is seen as getting a point across to the reader, there is still no difference. Meanwhile, poor Uther also has to face Lestat, who claims that well-written stories shouldn't have parts that you don't like, and, anyway, you can still skip ahead in a book. Uther, harried and cornered, is finally treed: "i can't define the difference exactly. The difference is what you make it out to be." Goaded for this by Lestat's "could you be more vague?" Uther wonderfully retorts with a realization that he is developing theory for himself: "I can be much more vague. But we are dealing with ideas and concepts here." The participants are tiring (as is the software, which clogs and slows and creates typos when the transcript grows too large) as they turn to what constitutes good writing. {*"good" writing (17)*} Lestat claims that every important thing he reads is good all the way through. This eventually leads to a consideration of the publishing choices that privilege "quality." Lestat wonders whether "with htext . . . the writer [can] afford to be 'less good' because peole can gloss over" the bad. Uther says "no, but as a reader you decide what is good or bad." The section, and the Interchange, end in a wisely comic sequence: "Lestat: how come ther is so much trash out then?//Uther: what appears in a published book is what many people decide is good. They don't put bad things in.//Danielle Steele: who's talking about me?//Uther: someone considers that 'trash' good. you just think it isn't good. //Uther: you don't publish something that nobody will read."

Descriptive summary—Comp Writers. {*preliminaries (12)*} After initial Interchange sessions early in the term, this class has largely done its group work in Storyspace. Thus, there is a great deal of recalling how to get started. Django's "This hypertext is very hard to understand" seems the kind of formal foray by which, even at midterm in a supposedly decentered, self-grading, collaboratively grouped class, students test your intentions and the reality of your conversion: Django knows I need research; he poses a research question. Given a ground, I build on it, framing my interests in two parts: "how this kind of writing environment changes your view of the writing process" and "what . . . you think hypertext is (if you think anything at all) [and how] . . . do you think about the parts of a writing . . . in a Storyspace?" The Stud and shazbat begin the thread I've key-worded as *penileHumor.* {*agenda hunt (22)*} Given the piles of fleece laid out by my agenda setting (and the dark fur of The Stud's antiagenda), a number of participants begin to hunt for a thread that will engage the group. Emotionally Ignorant considers collaboration, "determining whether or not this works for you depends on the group you are in and if they work well together." Denzel suggests that "the computer has helped my writing . . . [it] makes me aware of spelling, and form." Squirrel merges formal advantages and group advantages in a happy formulation: "it helps cuz there is more than one persons input on a paper instead of just one so no ones weekneses realy show up in the document." No one spins at the hypertext thread directly except the agenda-setter Django, who, despite his early weighing in regarding its difficulty, now "want[s] to know what hypertext is." Air Man—in an Interchange that at first orthographic glance would seem to confirm the suspicions of the deepest skeptic—eloquently evokes the spirit of hypertext and collaborative work in a fine insight about writing processes: "The exptation of this type of a writing inverment is to be able to write whatever you think when you think it and areange it when you nead to." When I realize now that each of these threads—formal advantages, group success, and symbolic manipulation (i.e., writing/arranging chunked though)—remains a matter of active inquiry through this whole Interchange session, I am surprised, unaware, either in the session itself or in my coding pass through this material, of these thematic threads. Other threads developed later are first spun here, especially in the invitations to joke telling. ZøЅó, the lab tutor assigned to this class, is also the only writer to take part in both the HT Writers' and Comp Writers' interchanges. Here he sees himself as my sidekick coteacher and colearner, and he establishes an in-joke thread that he and I will indulge ourselves in later in the session, "Blah blah hypertext blah blah interactive collaboration blah blah buzzword buzzword." {*re hypertext (30)*} The group ranges through a halting attempt to discuss my question about whether a hypertext writing environment really lets you choose what and when you want to read and write. Z joins in and (I think now) models typical Interchange style by narrating his experience with collaborative hypertext. I restate the concern as "What is different about writing with Storyspace? what is the same?" Since it follows only two interchanges later, Zoe's stunning response has likely been something she was composing through the series of jokes that followed my initial agenda setting. Her response is an example of what I've key-worded as *set pieces* and is, in fact, a lovely consideration of the question. Only now really, as I write this summary, do I discover this and add *hypertext* to the keywords for this response. Her dream meditation will resurface late in this session in a thread about reality:

> Zoe: When I lay down at night my mind slips back a few feet wondering about the day I just experienced and then jumps forward about ten miles. I can't keep up with it, sometimes I think I have insomina. I don't care if I ever sleep again as long as I have intense dreams, I could spend the rest of my life dreaming and wondering if I spelled *insomnia* correctly.

Air Man also continues to advance this question in his (orthographically more kosher) observation that "for a mixed up thinker who thinks randomly this is the way to write." Denzel likewise expands on his initial insight that escaping surface structure concerns frees one to consider real (rhetorical) structure. "When I write on paper it seems to be alot of jumble," Denzel says: "I not only have to deal with puncuation but with my own handwriting. But with the computer I have less to deal with so the structurs has a better foundation. And I find that people can read my writing with greater understanding." Finally, the aptly pseudonymed Aristotle, in an echo of Hawisher's question, asks, "Is this (discourse) part of hyperspace or storyspace?" {*changing writing (31)*} Of the many remarkable things about sassi's set piece, perhaps most remarkable is how little immediate effect it has upon the discourse (although it becomes a deep flowing stream that shapes following interchanges). Nonetheless, the set piece is a wonderful example of a fully formed improvisational composition. Beginning with a paragraph-long consideration of technology itself, how learning "that I don't have to be affraid of the computer"—"I can us it to do things that I would have never found out about [and] . . . to get intouch with the new world"—she then moves to another paragraph-long consideration of collaboration, which she values because "there are so many interesting people in here . . . [who] do really care if you succeed or if you just need their help they are there for you." Next she reframes Denzel's earlier point about how escaping surface structure concerns frees one to consider real structure, noting that "writing on computers, . . . you can move around paragraphs so that they might be more interesting to read. It is not as stresful as when you write longhand and then you find that you want to rearrange your paper, and that means that you have to rewrite your paper all over again." After a bow to the competitive advantage computers might give in a technological age, sassi closes with a query: "I'm glad that I took this class. . . . What about the rest of you???" Even her rhetorical question (unlikely to pass traditional classroom scrutiny) can be seen as beyond the essay, a punctuating query that announces sassi's return to the flow. Since the flow is constant, extended replies like hers get less attention than shorter screens do in electronic writings (Interchange, Storyspace, or e-mail). The group has unwittingly gone on to consider the issues sassi raises (the "mixed-up thought of hypertext, e.g., or Denzel's complaint that he "get[s] typing some times fast and mistype words"), but sassi must make her points in the flow again, as she did in the essay. She does so ably, telling Denzel: "Don't worry about misspelling words and putting things into improper sentence form that you can fix anytime . . ." and responding philosophically to the hypertext thread: "Life goes on any way you can get it to go and then some. . . ."

{*changing mind (17)*} The group turns from a consideration of changing writing to changing minds, prompted by yet another restatement of how computers free a writer. Liza advances the argument beyond surface structure concerns to rhetorical structure. "In story space," she says, "you can move things around without having to retype 50 times or haveing a dozen confusing line all over yoiur paper. It makes changing your mind much easier." Squirrel recognizes this shift immediately: "I know that

I don't think in order sometimes I dont even make sence when I talk." Z suggests to Liza that, once she understands how to manipulate ideas in Storyspace, "if you use links . . . [you] make your mind multiple (hurm, cubist writing?)." *{what's a link (54)}* At this point, in response to my earlier query "Do any of the habits you develop writing on computers stay with you when you write on paper?" Denzel makes an agenda-setting claim that in "the few weeks I have been working on this computer I have noticed a stronger structure in the 'paper' writing." This develops into a thread about structure in both writing and mind. Air Man claims that "people must think in order—that's why they get stuck and can't write." In the immediately following Interchange (which thus likely comes as a happy match of concerns rather than a direct response) Jon Cakes disagrees: "I dont think that anyone thinks in order or structured form. . . . You just see where the writing is going and help it along. . . . [It] goes where it wants to you have to control it and structure it to keep it from becoming just a written stream of conciousness the computer really helps do this." How it does so becomes a matter of inquiry, as Liza and Z pursue a dialogue about the technique of hypertext linking. "With hypertext," says the hypertext writer Z, "I am more inclined to write something and . . . if I want to change it I'll write something else and cross-link it." Pressed by Liza, Socrates, and others to "actually explain," Z says: "think of making a path frome piece of text to another . . . [and] find something (a word, a thought) that looks like a place to leave to somewhere else, and then link away" (see Landow 1989). Much later in this section Denzel expands on what stronger structure means and appropriates this language: "The writing is easer to follow. instead of incompleat sentences my stories have place to go . . . you can see the mistakes that may make it hard for the reader to follow." Much of this thread is filled with noise from the joke crowd and penile humor one-noters. Even sassi, her set piece done, contributes blond jokes. hat trick's questioning of the discourse, "Why do we have grade school chatter on collage level communications?" seems ironic since s/he's mostly told jokes here. Even so, the jokes and the thread commingle in Socrates' question, "Was'nt Link part of the 'Mod Squad'?"

{lost in the net (53)} When 90210 says "I cannot follow this," it is as if it is a signal for the group to release into drift. It is interesting that this drift comes at about the same midway point as the hypertext writers' drift also did. Even so, sprinkled through the drift, so separated that on first and second pass it is difficult to see—and impossible to see in the Interchange itself, in which I went from calling like a voice in the wilderness for responses to suggesting that we call it a day—there emerges the first twistings of the thread that follows. Socrates' claim "that this course has changed the way I feel about writing . . . from more of a personal activity to a group experience" is affirmed by Denzel's claim that it "has been . . . fun working in this class and . . . finding out what people think of my writing." sassi challenges Socrates, "Is that good or bad??? going from personal to group writing??" Socrates replies in terms of the rhetorical situation: "It's good because it has let me explore a whole new way of writing. Before when I wrote, I wrote more for myself, now I must think of how others may interpert my writing." The appearance of "12-incher" with the obvious question "what's up?" signals a deliberate interference and agenda setting from the penile humor boys, who may be distressed by the emerging threads. This kind of interference continues for the remaining sections. *{group thread (35)}* A self-organizing, self-referencing system, the drift suddenly dissipates into a clear channel discussion of the value of collaborative writing and the relationships of group and individual writer. Goofy locates the value of collaboration in the actual "working together as a group in writing," which, he says, "is alot of fun, if you let it be. . . . when you don't have anything to say, someone else can step in and put something down." Jon Cakes disputes this view and argues for the traditional romantic role of the solitary author, "writing as more of a personal experience where it's just you interacting with your work rather than having people muddle your creativity with their own mental trash." Even so, he finds a value for the group in the editorial, reader-focused part of the writing process: "It's nice to have people to correspond with and get opinions from, but I really don't like the group thing until the preliminary writing is done." I challenge this, asking whether a writer needs to have the readers' responses in mind even as she

writes in private, and a (literally) nameless participant responds, "Nothing is real unless it is observed." Socrates (!) affirms the "real in the beholder's mind." The more pragmatic Squirrel again picks up on how computers free a writer to consider rhetorical structure. "I have actualy enjoyed [this class]," squirrel says, "because the preshure wasnt on the actual structer and spelling but more on the colaberative writein." Jousting with Jon Cakes, sassi invents social construction for herself by locating the personal in the group: "The personal is just that personal... but the interaction with the group gives you other ideas to think about... as well. Wouldn't you say so??" Jon Cakes won't say so but nonetheless discovers social construction and reader response for *himself*, albeit in the obverse. "My writing usually turns out Lame when I think of what the response will be to it," Jon Cakes says: "When you think about the readers response too much you start to contaminate your writing with your view of what people expect of your writing." Denzel remains consistent in his focus on structure as the context (he calls it situation) for invention, putting himself squarely in Jon Cakes's solitary, romantic, self-sufficient (male?) camp. "When I write I think of the situation I have got Myself into and the way to get out sucessfuly," he says. "If I think of my readers I tend to lose that special something that my writing needs. Then I look over what I have said and then detirmine if I have compleated what I was hoping to achieve in the writing." {*what's a rube (34)*} Once again (it is late in the hour) the group releases into drift. Air Man recapitulates the discussion about group, self, and the real in a proposition that is either a paradox about electronic writing or the victim of a typo: "If nothing is real unless it is observed then our writing is real as long as it is in our files and not read." As the group prepares to leave, squirrel sums up life ("smetimes you have to be serious and you need to know when you can and don't need to be") and writing ("I prefer eletronic writeing [because of] the ability to change the document without starting over and the spellchecker"), but the final word is Denzel's: "This class has taught me... writing that I have never thought of before.... I have notice that writing is a mind filled thing..."

Notes

Introduction

For texts cited in this introduction see following chapters and Byrd 1991; Deleuze and Guattari 1983; Douglas 1991; Guyer 1992; Haraway 1991; Hayles 1992; Kristeva 1991; and Landow 1992.

Siren Shapes

This article first appeared in *Academic Computing* 3, no. 4 (1988): 10–14, 37–42, and is reprinted by permission.

What the Fish Lady Saw

This essay was among a number distributed to participants in an all-day workshop preceding the 1992 CCCC meeting. An annotated table of key words, which appeared with the original essay, is not included here.

Storyspace Hypertext Writing Environment for the Macintosh computer is published by Eastgate Systems, P.O. Box 1307, Cambridge, MA 02238, and was developed by Jay Bolter, John B. Smith, and myself. Link Apprentice is an unpublished, experimental suite of tools, including linkplots, under development by Eastgate Systems Chief Research Scientist, Mark Bernstein. The Discourse suite of programs, including Interchange, is published for MS-DOS and Macintosh computers by The Daedalus Group, Inc., 300 E. Huntland Dr., Suite 222, Austin, TX 78752.

Interstitial: Everyone's Story Goes On without Us

For information about Hi-Pitched voices, a collaborative hypertext writing project for women, write <caroway@aol.com>.

1. The more theoretical quotations above are from Moulthrop's "You Say You Want a Revolution? Hypertext and the Laws of Media"; the others are from *Victory Garden.*

"So Much Time, so Little to Do"

This essay was originally presented as the keynote address at the Sixth Conference on Computers and Writing, Austin, Tex., 17 May 1990.

A Memphite Topography

This essay was originally presented in a different form as a keynote address at the Language: Future Tense Conference sponsored by the National Endowment for the Humanities (NEH), Memphis State University, and the Tennessee Department of Education at Memphis, Tenn., 16 September 1990.

New Teaching

This essay was published in *Computers and Composition* 9, no. 2 (April 1992); and was originally presented at the CCCC in Boston, 1991.
Carolyn Guyer's sense of "preservation" as a constructive action informs the discussion of the role of learning manager (a term she dislikes) here.

Interstitial: Silicon Valley Maoists and Ohio Zen

In memory of Paul Petry.

Selfish Interaction

This essay was published in *The McGraw-Hill Handbook of Hypertext and Hypermedia,* Emily Berk and and Joseph Devlin, ed. (New York: McGraw-Hill, 1991).

1. By far the most interesting story generation programs have emerged from the Yale Artificial Intelligence Lab. James Meehan's TALE-SPIN (*The Metanovel: Writing Stories by Computer.* Ph.D. diss., Research Report no. 74, Computer Science Department, Yale University) builds Aesopian stories in natural language according to a conceptual representation of the story-world. Natalie Dehn's (1981, 1989) AUTHOR was an attempt to generate a story based upon the intentions of the computer "author." Dehn's work on STARSHIP, as yet unpublished, moved even closer to genuine interaction, creating a science fiction world in which stories changed according to the program's perception of the reader's comprehension of the story as it developed, based upon comprehension questions generated by the program. Much of my thought on interaction and multiple fiction is indebted to Dr. Dehn (currently at Martin Marietta) and, in a more oblique way, to Roger Schank, who was good enough to invite me to visit the Yale Lab, though he wasn't sure whether I would profit from it. Schank seemed alone, in the increasingly applications-oriented world of AI, in his commitment to understanding human thought, learning, and invention.
2. According to the *New York Times* (5 December 1985, 2), "Mom is a Jewish Mother computer personality," and Murray "is a cartoon computer friend... conversing for hours with whomever is at the keyboard." The program was created by Yakov Kirshen and was marketed by Antic for the Atari 520ST. Eliza, of course, refers to Joseph Weizenbaum's widely adapted and nearly legendary computer program. (See Weizenbaum 1966).
3. Early development of Storyspace was supported in part by a grant from The John and Mary H. Markle Foundation.

4. Eco 1979, 9.

5. Douglas R. Hofstadter argues a similar point: "And when a novelist simultaneously entertains a number of ways of extending a story, are the characters not, to speak metaphorically, in a mental superposition of states? If the novel never gets set to paper, perhaps the split characters can continue to evolve their multiple stories in their author's brain. Furthermore, it would even seem strange to ask which story is the *genuine* version. All the worlds are equally genuine" (1985, 472). My notion is that this state holds true even when the novel is set to paper.

6. It is a shame that this extraordinary novel (1966) needs to be footnoted, but so be it. The author's "Table of Instructions," begins: "In its own way, this book consists of many books, but two books above all. The first can be read in normal fashion and it ends with Chapter 56. . . . The second should be read beginning with Chapter 73 and then following the sequence indicated . . ." The sequence moves back and forth through "new" (post-chap. 56) material and old. Cortázar is best known as the author of *Blow-up* (1963), which, of course, was the inspiration for Antonioni's film of the same name.

7. The so-called Gorman-Gilbert schema for *Ulysses* provides graphic substantiation of such pre-processing, especially since Joyce originally circulated it on a not-for-publication basis among friends and potential critics. See the appendix to Ellmann 1972 (186–87), for concise history and reproduction of the schema.

8. Eco 1979, 62 (his italics).

9. Ibid., 63 (his italics).

10. *Some* readers, of course, do. The Baker Street Irregulars are perhaps best known, although science fiction fans seem even more devoted to creating an alternate canon. The Friends of Darkover, "a non-sectarian, non-sexist, and non-profit group of science fiction and fantasy lovers... has come into being with no purpose except the discussion of secondary universes, primarily Darkover" (statement of purpose, *The Keeper's Price*, by Marian Zimmer Bradley and the Friends of Darkover, DAW Books no. 373, New York, 1980). Bradley mentions similar alternate worlds created by Star Trek and Tolkien fans. Friends and colleagues of mine report similar alternate versions created by fans of the "Dr. Who" television program and readers of *The Witch World of Andre Norton*. Bradley notes that the majority of alternate Darkover stories are written by women and suggests, much in the same vein as Sherry Turkle, that women, who are trained as children by role-playing rather than fantasy, tend to feel more comfortable at first in someone else's world.

11. They might also use Storytree, a simple and quite fascinating structure editor, published by Scholastic Software.

12. See note 7. (This note is, of course, a Barthian interaction, à la *Lost in the Funhouse*.)

13. This and all following quotations, with the hopefully self-evident exception of the last line from "Big Two-Hearted River," are from Eco 1979, 86 (his italics).

The Geography of the Word

This essay was originally published in the *Bulletin of Science, Technology, and Society (STS)* 7, no. 4 (1987) and is reprinted with permission.

1. Hofstadter and Dennett 1981, 385.
2. Olson 1947, 11.
3. Sauer 1963, 391.
4. I owe both this insight, as well as the specific reference to Derrida, to Jay Bolter, who, as an incidental aspect of our continuing collaboration, shared them with me and who, as an integral aspect of our collaboration, has stimulated me to think about the meaning and evolution of writing spaces.
5. Derrida 1979.
6. Storyspace was created by Jay Bolter and myself (in conjunction with John B. Smith, also of the University of North Carolina [UNC]), and early development was partially supported by a grant from The Markle Foundation. Storyspace is a trademark of Eastgate Systems, Cambridge, Mass.

   Structure editors are known variously as "idea processors," "hypertext systems," and "outline processors," although these terms are not really interchangeable, nor do they describe the differences among computer programs and systems that act as structure editors. Storyspace is, in fact, a hypertext structure editor; i.e., it enables nonlinear writing and reading with user-controlled links. Other microcomputer hypertext systems include Guide and Hypercard; microcomputer outline processors include MORE, its progenitor Thinktank, Agenda, and Acta, among others; the list of programs that claim to be idea processors is too long to enumerate here.
7. Sauer 1963, 393.
8. Barthes 1967, 13.
9. Zukav 1979, 212.
10. Ibid., 220.
11. Ibid., 220–21, from Louis de Broglie, "A General Survey of the Scientific Work of Albert Einstein," in *Albert Einstein, Philosopher-Scientist*, vol. 1, ed. Paul Schilp (New York: Harper and Row, 1949).
12. I am grateful to Roger Schank for his generosity in inviting me to the Yale Lab as a Visiting Fellow while I was on sabbatical during the 1984–85 academic year, and to Natalie Dehn, now of the Martin Marietta Corporation, for both instigating that invitation and consistently challenging my assumptions about creativity and the writing process.

I should also note that my characterizations of their work here are neither the "Authorized Version," nor even "The Revised Standard Version," but rather what one hazards in letting novelists hang around ideas.

13. Roger Schank and Robert Abelson (1977), p.11.
14. Ibid., 17, 3–7, 41.
15. Schank 1982.
16. Sauer 1963, 321.
17. Ibid., 322.
18. Ibid., 324.
19. Ibid., 325.
20. Natalie Dehn, personal communication, November 1983.
21. Dehn 1981.
22. Sauer 1963, 400.
23. Ibid., 401.
24. Ibid., 406.
25. Ibid., 403.
26. The concept of the "dribble file" is the creation of Alan Kay, a research fellow with Apple Computer and, truly, a seminal force in the theory and practice of human-machine interfaces. Much of what we have come to expect in computer interfaces is in some way influenced by his thought.
27. Bachelard 1965; quoted in Bachelard 1971, 52.

"The Ends of Print Culture"

This essay is adapted from a talk given originally at the Computers and Human Conversation Conference, Lewis and Clark College, Portland, Ore., 16 March 1991.

Interstitial: Artists' Statements—Giving Way(s) before the Touch

The first statement is reprinted from the introductory screens of *afternoon, a story,* (1987) published by Eastgate Systems (1990). The second statement first appeared in "Words on Works," which pioneer electronic writer Judy Malloy edits for the *Leonardo Electronic News* published by the International Society for the Arts, Sciences and Technology.

Hypertext Narrative

This essay, published in *Perforations* (3 [Spring–Summer 1992], guest edited by Richard Gess, under the editorship of Robert Cheatam), was originally presented in different form as part of the panel "Hypertext, Narrative, and Consciousness" at the ACM Hypertext '89 meeting, Pittsburgh, Pa., 6 November 1989.

What Happens as We Go?

The first part of this essay is from a talk given in different form at the Museum of Modern Art, as part of the "Technology in the 90s" lecture series, organized by Barbara London, New York, on 26 April 1993.

The Momentary Advantage of Our Awkwardness

1. See Neruda's poem "Pact (Sonata)":

> By now sometimes it is not possible
> to win except by falling,
> by now it is not possible to tremble between
> two beings, to touch the flower of the river

2. Cf. Laurie Anderson, United States Live (I–IV), audio CD, Burbank and New York: Warner Bros. Records, Inc., 1984:

> You're walking. And you don't always realize it,
> but you're always falling.
> With each step, you fall forward slightly.
> And then catch yourself from falling.
> Over and over, you're falling.
> And then catching yourself from falling.
> And this is how you can be walking and falling
> at the same time.

A Feel for Prose

Published in *Writing on the Edge* 3, no. 1 (1992) from a talk given as part of "Wednesdays at 4 PLUS" series, State University of New York at Buffalo, 13 February 1992.

1. "As their bodies jostle in a bus . . . when I am rocked by the roads against any of them–kids, women, men–. . . touch is in no sense anything else but the natural law of flesh . . . " (Olson 1951, 57).

2. In Milorad Pavič's novel *Dictionary of the Khazars* (1988): "One of the envoys had the Khazars' history and topography tattooed on his body" (73) "and lived . . . like a living encyclopedia of the Khazars, on money earned by standing quietly through long nights. He would keep his vigil, his gaze fixed on the Bosporus' silver treetops, which resembled puffs of smoke. While he stood, Greek and other scribes would copy the Khazar history from his back and thighs into their books" (77).

3. "In striated space, one closes off a surface and 'allocates' it according to determinate intervals, assigned breaks; in the smooth, one 'distributes' oneself in an open space, according to frequencies and in the course of one's crossing (*logos* and *nomos*)" (Deleuze and Guattari 1983a, 481).

4. "The possibilities which the work's openness makes available always work within a given field of relations . . . we can say that the *work in movement* is the possibility of numerous personal interventions, but it is not an

amorphous invitation to indiscriminate participation. The invitations offers the performer the chance of oriented insertion into something which always remains the world intended by the author" (Eco 1979, 62).

5. "In all claims to the story... there is muteness. The writer as witness, speaking the stories, is a lie, a liberal bourgeois lie. Because the speech is the writer's speech, and each word of the writer robs the witnessed of their own voice, muting them" (Mouré 1989, 84).

6. From the song "Once in a Lifetime," by David Byrne, Chris Frantz, Jerry Harrison, Tina Weymouth, and Brian Eno (*Stop Making Sense*, Talking Heads, 1983).

7. "But then making an effort we may withdraw attention from the garden; and by retracting the ocular ray, we may fixate it upon the glass. Then the garden will disappear in our eyes and we will see instead only some confused masses of color which seem to stick to the glass. Consequently to see the garden and to see the glass in the window-pane are two incompatible operations" (Ortega y Gasset 1956, 67).

8. "Some early Greek inscriptions were written in a style called boustrophedon... in which the line ran left to right, bent around, and then continued from right to left with individual letters also drawn backwards... the technique was perfectly linear" (Bolter 1991, 108).

9. "Assemblages are passional, they are compositions of desire. Desire has nothing to do with a natural or spontaneous determination; there is no desire but assembling, assembled desire. The rationality, the efficiency, of an assemblage does not exist without the passions the assemblage brings into play, without the desires that constitute it as much as it constitutes them" (Deleuze and Guattari 1983a, 399).

10. "Sound exists only when it is going out of existence. It is not simply perishable but essentially evanescent, and it is sensed as evanescent. When I pronounce the word 'permanence,' by the time I get to the '–nence,' the 'perma–' is gone, and has to be gone... There is no way to stop sound and have sound" (Ong 1982, 32).

11. "An oral culture may well ask in a riddle why oaks are sturdy, but it does so to assure you that they are, to keep the aggregate intact, not really to cast doubt on the attribution.... Traditional expressions in oral cultures must not be dismantled: it has been hard work getting them together over the generations, and there is nowhere outside the mind to store them" (Ong 1982, 39). So too for electronic texts there is nowhere outside the mind or machine to store them.

12. "To create a music of contrasting characterizations, so that you can have not only this monoplanar or dyadic movement to characterization, framing the frame, but that you can have lots of different angles in composition so that the whole sounding of the various characterizations gets heard and made palpable" (C. Bernstein 1986b, 446).

13. "Electronic writing is both a visual and verbal description. It is not the writing of a place, but rather a writing *with* places, spatially realized topics. Topographic writing challenges the idea that writing should be merely the servant of spoken language. The writer and reader can create and examine signs and structures on the computer screen that have no easy equivalent in speech" (Bolter 1991, 25).

14. "Exploratory hypertext, which most often occurs in read-only form, allows readers to control the transformation of a defined body of material. Its demise is into the stylized abstraction of orthodoxy, the second of Latour's alternatives, and as such is inherently dangerous. To the extent that exploratory hypertexts give us the illusion of control, we are apt to believe too much in them, and so risk sealing others' power over us through our tacit practice. Control itself can be made to seem a product" (Joyce 1990, in this volume).

Works Cited

Adams, Douglas. 1979. *The Hitchhiker's Guide to the Galaxy.* New York: Harmony Books.

Althusser, Louis. 1971. "Ideology and the State." In *Lenin, Philosophy and Other Essays.* Translated by Ben Brewster. New York and London: Monthly Review Press.

Anderson, Laurie. 1984. *United States Live,* vols. 1–4. Audio CD. Burbank and New York: Warner Bros. Records.

Bachelard, Gaston. 1971. *On Poetic Imagination and Reverie.* Selected and Translated by Collette Gaudin. New York: Bobbs-Merrill.

——. 1965. "*La Terre et les reveries du repos.*" Paris: Librairie José Corti.

Balestri, Diane Pelkus. 1988. "Softcopy and Hard: Wordprocessing and Writing Process." *Academic Computing,* 2, no. 5 (February 1988): 14–17 and 41–44.

Barthes, Roland. 1967. *Writing Degree Zero.* Translated by Annette Lavers and Colin Smith. Boston: Beacon Press.

Batson, Trent. 1989. "Teaching in Networked Classrooms." In *Computers in English and the Language Arts,* edited by C. L. Selfe, D. Rodrigues, and W. R. Oates, 247–55. Urbana, Ill.: National Council of Teachers of English.

——. 1988. "The ENFI Project: A Networked Classroom Approach to Writing Instruction." *Academic Computing* 3, no. 4: 32–33, 55–56.

Becker, Howard S. 1986. *Writing for Social Scientists: How to Start and Finish Your Thesis, Book, or Article* (with a chapter by Pamela Richards). Chicago: University of Chicago Press.

——. 1982. *Art Worlds.* Berkeley: University of California Press.

——. 1973. *Outsiders: Studies in the Sociology of Deviance.* New York: Free Press.

Beeman, William O., et al. 1987. "Hypertext and Pluralism: From Lineal to Non-Lineal Thinking." *Proceedings of Hypertext '87,* Association for Computing Machinery (ACM), Chapel Hill.

Berk, Emily, and Joseph Devlin, eds. 1991. *McGraw-Hill Handbook of Hypertext and Hypermedia.* New York: McGraw-Hill.

Bernstein, Charles. 1991. "PLAY IT AGAIN, PAC-MAN." *Postmodern Culture* 2, no. 1 (September).

——. 1986a. Interview with Tom Beckett. *Content's Dream, Essays, 1975–1984*, 385–411. Los Angeles: Sun and Moon Press.

——. 1986b. "Characterization." *Content's Dream, Essays, 1975–1984*, 428–62. Los Angeles: Sun and Moon Press.

Bernstein, Mark. 1991a. The Navigation Problem Reconsidered. *McGraw-Hill Handbook of Hypertext and Hypermedia,* edited by Emily Berk and Joseph Devlin. New York: McGraw-Hill.

Bernstein, Mark, Jay David Bolter, Michael Joyce, and Elli Mylonas. 1991b. "Architectures for Volatile Hypertext." *Proceedings of Hypertext '91* ACM, San Antonio.

Bey, Hakim. 1991. T. A. Z. *The Temporary Autonomous Zone: Ontological Anarchy, Poetic Terrorism.* Brooklyn: Autonomedia.

Bøgh Andersen, Peter. 1994. "The Force Dynamics of Interactive Systems: Towards a Computer Semiotics." *Semiotica.* Forthcoming.

Bolter, Jay David. 1991. *Writing Space: The Computer, Hypertext, and the History of Writing.* Hillsdale, N.J.: Lawrence Erlbaum and Associates.

——. 1984. *Turing's Man: Western Culture in the Computer Age.* Chapel Hill: University of North Carolina Press.

Bolter, Jay David, Michael Joyce and John B. Smith. 1990. *Storyspace: Hypertext Writing Environment for the Macintosh.* Computer software. Cambridge, Mass.: Eastgate Systems.

Brand, Stewart. 1987. *The Media Lab: Inventing the Future at MIT.* New York: Viking.

Brown, P. J. 1987. "Turning Ideas into Products: The Guide System." *Proceedings of Hypertext '87.* ACM, Chapel Hill.

Bruffee, Kenneth A. 1990. "Collaborative Learning: Language and the Authority of Teachers." Paper presented at Conference of College Composition and Communication, Chicago.

Bruner, Jerome. 1986. *Actual Minds, Possible Worlds.* Cambridge, Mass.: Harvard University Press.

Bush, Vannevar. 1945. "As We May Think." *Atlantic Monthly* 176, July, 101–8.

Byrd, Don. 1991. "Cyberspace and Proprioceptive Coherence." Paper presented at the Second International Conference on Cyberspace, Santa Cruz, Calif., 20 April.

Calvino, Italo. 1988. *Six Memos for the Next Millennium.* Cambridge, Mass.: Harvard University Press.

——. 1982. *The Uses of Literature.* Translated by Patrick Creigh. New York: Harcourt Brace Jovanovich.

——. 1968. "All at one point." *Cosmicomics.* Torino: Guilio Editore s.p.a., 1965. English translation by William Weaver. New York: Harcourt Brace Jovanovich and Jonathan Cape.

Charney, Davida. 1994. "The Effect of Hypertext on Processes of Reading and Writing." In *Literacy and Computers,* edited by Susan Hillgloss and Cynthia Selfe. New York: MLA.

Cixous, Hélène. 1991. *Coming to Writing and Other Essays.* Edited by Deborah Jenson. Cambridge, Mass.: Harvard University Press.

Clifford, J., and G. Marcus. 1986. *Writing Culture: The Poetics and Politics of Culture.* Berkeley: University of California Press.

Clifton, Lucille. 1987. "To Merle." *Good Woman: Poems and a Memoir, 1969–1980.* Brockport, N.Y.: Boa Editions.

Conklin, Jeffrey. 1987. "Hypertext: An Introduction and Survey." *IEEE Computer* 20, no. 9 (September): 17–41.

Cooper, Marilyn M., and Cindy L. Selfe. 1990. "Computer Conferences and Learning: Authority, Resistance and Internally Persuasive Discourse." *College English* 52, no. 8 (Fall): 847–69.

Coover, Robert. 1993. "Hyperfiction: Novels for the Computer." *New York Times Book Review,* 29 August, 1, 8–12.

——. 1992. "The End of Books." *New York Times Book Review,* 21 June, 1, 23, 24–25.

——. 1969. *Pricksongs and Descants.* New York: Plume.

Cortázar, Julio. 1966. *Hopscotch.* Translated by Gregory Rabassa. New York: Random House.

Creeley, Robert. 1970. "I'm Given to Write Poems." *A Quick Graph: Collected Notes and Essays,* 61–72. San Francisco: Four Seasons Foundation.

Dehn, Natalie Jane. 1989. "Computer Story-Writing: The Role of Deconstructive and Dynamic Memory." YALEU/DCS/TR792, Yale University, New Haven.

——. 1981. Story Generation after Tail-Spin. *Proceedings of the Seventh Annual Joint Conference on Artificial Intelligence,* Vancouver, B.C.

Deleuze, Gilles, and Félix Guattari. 1987. "The Desiring-Machines." *Anti-Oedipus: Capitalism and Schizophrenia.* Minneapolis: University of Minnesota Press.

——. 1983a. "The Smooth and the Striated." *A Thousand Plateaus: Capitalism and Schizophrenia.* Minneapolis: University of Minnesota Press.

——. 1983b. "Treatise on Nomadology–The War Machine." *A Thousand Plateaus: Capitalism and Schizophrenia.* Minneapolis: University of Minnesota Press.

Derrida, Jacques. 1979. "Living On." Translated by James Hulbart. In *Deconstruction and Criticism,* edited by Geoffrey Hartman, 75–76. A Continuum Book. New York: Seabury Press.

Douglas, Jane Yellowlees. 1994. *Print Pathways and Interactive Labyrinths.* Baltimore: Johns Hopkins University Press. Forthcoming.

——. 1991. "The Act of Reading: The WOE Beginners' Guide to Dissection." *Writing on the Edge* 2, no. 2 (June): 112–25.

——. 1987. "Beyond Orality and Literacy." Paper presented at the annual meeting of the International Association for Computers in Education, New Orleans.

Eco, Umberto. 1989. *The Open Work.* Translated by Anna Cancogni. Cambridge, Mass.: Harvard University Press.

——. 1979. *The Role of the Reader.* Bloomington: Indiana University Press.

Ellmann, Richard. 1972. *Ulysses on the Liffey.* New York and London: Oxford University Press.

Engelbart, Douglas. 1963. "A Conceptual Framework for the Augmentation of Man's Intellect." In *Vistas in Information Handling,* edited by Paul W. Howerton and Donald C. Weeks, 1–29. Washington, D.C.: Spartan Books.

Feldman, C., and J. Wertsch. 1976. "Context Dependent Properties of Teachers' Speech." *Youth and Society* 8: 227–58.

Friedrich, O. 1989. *Glenn Gould: A Life and Variations.* New York: Random House.

Gibson, William. 1987. *Burning Chrome.* New York: Ace.

——. 1984. *Neuromancer.* New York: Ace.

Glushko, Robert. 1989. "Design Issues for Multi-document Hypertexts." *Proceedings of Hypertext '89,* ACM, Baltimore.

Gould, Glenn. 1966. "The Prospects of Recording." *High Fidelity* (April).

——. 1964. "Strauss and the Electronic Future." *Saturday Review,* 30 May.

Graves, Robert. 1955. *The Greek Myths.* Baltimore: Penguin.

Guyer, Carolyn. 1993a. "Artist's Statement." *Leonardo Electronic News,* Works on Words issue 2, no. 11 (15 November).

——. 1993b. *Quibbling.* Computer software. Cambridge, Mass.: Eastgate Systems.

——. 1992. "Buzz-Daze Jazz and the Quotidian Stream." Paper presented at the Modern Language Association (MLA) Convention, New York City.

Guyer, Carolyn, and Martha Petry. 1991a. *Izme Pass,* a collaborative hyperfiction. *Writing on the Edge* 2, no. 2. Bound-in computer disk, University of California at Davis.

——. 1991b. "Notes for Izme Pass Exposé." *Writing on the Edge* 2, no. 2. University of California at Davis.

Haas, Christina. 1988. "Planning in Writing: The Influence of Writing Tools." Paper presented at Conference on College Composition and Communication (CCCC), St. Louis, March.

Halasz, Frank. 1987. "Reflections on Notecards: Seven Issues for the Next Generation of Hypermedia Systems." *Proceedings of Hypertext '87.* ACM, Chapel Hill.

Haraway, Donna. 1991. "A Cyborg Manifesto: Science, Technology, and Socialist-Feminism in the Late Twentieth Century." *Simians, Cyborgs and Women.* New York: Routledge.

Harpold, Terence. 1994. "Conclusions." In *Hypertext and Literary Theory,* edited by George Landow. Baltimore: John's Hopkins University Press. Forthcoming.

——. 1991. "Threnody: Psychoanalytic Digressions on the Subject of Hypertexts." In *Hypermedia and Literary Studies,* edited by Paul Delany and George Landow, 171–84. Cambridge, Mass.: MIT Press.

——. 1990. "The Grotesque Corpus: Hypertext as Carnival." Paper presented at Computers and Writing Conference, Austin, Tex.

Hawisher, Gail, and Cindy L. Selfe. 1991a. Letter from the Editors. *Computers and Composition* 8, no. 3.

——. 1991b. "The Rhetoric of Technology and the Electronic Writing Class," *College Composition and Communication* 42, no. 1: 55–65.

Hawisher, Gail E., and Cynthia L. Selfe, eds. 1991. *Evolving Perspectives on Computers and Composition Studies: Questions for the 1990s.* Urbana, Ill., and Houghton, Mich.: NCTE Press and Computers and Writing.

Hayles, N. Katherine. 1992. "Gender Encoding in Fluid Mechanics: Masculine Channels and Feminine Flows." *differences* 4, no. 2: 16–43.

Heim, Michael. 1987. *Electronic Writing: A Philosophical Study of Word Processing.* New Haven and London: Yale University Press.

Herzfeld, Michael. 1991. *A Place in History: Social and Monumental Time in a Cretan Town.* Princeton: Princeton University Press.

Hofstadter, Douglas R. 1982. *Metamagical Themas.* New York: Basic Books.

Hofstadter, Douglas R., and Daniel C. Dennett. 1981. *The Mind's I.* New York: Basic Books.

Howe, Susan. 1985. *My Emily Dickinson.* Berkeley: North Atlantic Books.

Irigaray, Luce. 1993. *je, tous, nous: Toward a Culture of Difference.* Translated by Alison Martin. New York: Routledge.

Johnson-Eilola, Johndan. 1991. "Trying to See the Garden: Interdisciplinary Perspectives on Hypertext Use in Composition Instruction." *Writing on the Edge* 2, no. 2.

Jonassen, David H. 1988. "Designing Structured Hypertext and Structuring Access to Hypertext." Paper presented at the IBM ACIS Conference. Dallas, June.

Joyce, Michael. 1992a. "A Feel for Prose: Interstitial Links and the Contours of Hypertext." *Writing on the Edge* 4, no. 1.

——. 1992b. "New Teaching: Toward a Pedagogy for a New Cosmology." *Computers and Composition* 9, no. 3.

——. 1991b. "Notes Toward an Unwritten Nonlinear Electronic Text: The Ends of Print Culture." *Postmodern Culture* 2, no. 1 (elec. journal <pmc@unity.ncsu.edu>).

——. 1991a. "Selfish Interaction." In *The Hypertext/Hypermedia Handbook,* edited by Emily Berk and Joseph Devlin. New York: McGraw-Hill.

——. 1991c. "Storyspace as a Hypertext System for Writers and Readers of Varying Ability." *Proceedings of Hypertext '91*. ACM, San Antonio.

——. [1987] 1990a. *afternoon, a story.* Computer disk. Cambridge, Mass.: Eastgate Press.

——. 1990b. "So Much Time, So Little to Do: Empowering Silence and the Electric Book." Keynote address, Sixth Conference on Computers and Writing, Austin, Tex., 17 May.

——. 1988. "Siren Shapes: Exploratory and Constructive Hypertexts." *Academic Computing* 3, no. 4: 10–14, 37–42.

——. 1978. "Teaching Composition in a New Elizabethan Age." *College English* 39, no. 8: 894–903.

Joyce, Rosemary A. 1993. "Uses of History: Notes on Mesoamerica." *Symbols*. the Peabody Museum newsletter, Harvard University.

Kaplan, Nancy. 1991. "Reading and Writing the Word: Ideology, Technology, and the Future of Writing Instruction." In *Evolving Perspectives on Computers and Composition Studies: Questions for the 1990s,* edited by Gail E. Hawisher and Cynthia L. Selfe. Urbana, Ill., and Houghton, Mich.: NCTE Press and Computers and Writing.

Kaplan, Nancy, and Stuart Moulthrop. 1990. "Messages and Machines: Other Ways of Seeing." *Computers and Composition.* Urbana, Ill., and Houghton, Mich.

Kristeva, Julia. 1991. *Strangers to Ourselves.* Translated by Leon S. Roudiez. New York: Columbia University Press.

——. 1974. "Polylogue." *Tel Quel* 57. (Also in *Polylogue.* 1977. Paris: Editions du Seuil.)

Landow, George P. 1992. *Hypertext: The Convergence of Contemporary Critical Theory and Technology.* Baltimore: Johns Hopkins University Press.

——. 1989. "The Rhetoric of Hypermedia: Some Rules for Authors." *Journal of Computing in Higher Education* 1, no. 1: 39–64.

Landow, George P., and Jon Lanstedt. 1992. *In Memoriam Web.* Computer disk. Cambridge, Mass.: Eastgate Systems.

Lanham, Richard. 1989. "The Electronic Word: Literary Study and the Digital Revolution." *New Literary History 20,* no. 2: 265–90.

Latour, Bruno. 1989. *Science in Action.* Cambridge, Mass.: Harvard University Press.

Lombardo, Sergio. 1979. SPECCHIO TACHISTOSCOPICO CON STIMULAZIONE A SOGNARE, installation.

McCorduck, Pamela. 1991. *Aaron's Code: Meta-Art, Artificial Intelligence, and the Work of Harold Cohen.* New York: W. H. Freeman.

McDaid, John. 1993. *Uncle Buddy's Phantom Funhouse.* Computer software. Cambridge, Mass.: Eastgate Systems.

——. 1991. "Toward an Ecology of Hypermedia." In *Evolving Perspectives on Computers and Composition Studies: Questions for the 1990s,* edited by Gail E. Hawisher and Cynthia L. Selfe. Urbana, Ill., and Houghton, Mich.: NCTE Press and Computers and Writing.

——. 1990. "Hypermedia Composition and Consciousness." Paper presented at Computers and Writing Conference, Austin, Tex.

McLuhan, H. Marshall, and E. McLuhan. 1988. *Laws of Media: The New Science.* Toronto: University of Toronto Press.

Markoff, John. 1992. "Beyond the Personal Computer: Apple's Promised Land." *New York Times,* 15 November, sec. 3, pp. 1, 10.

Moulthrop, Stuart. 1991a. "Beyond the Electronic Book: A Critique of Hypertext Rhetoric." *Proceedings of Hypertext '91.* ACM, San Antonio.

——. 1991b. "CHAOS." Hyperfiction computer program. Atlanta.

——. 1991c. *Victory Garden.* Hyperfiction computer program. Cambridge, Mass.: Eastgate Systems.

——. 1991d. "You Say You Want a Revolution? Hypertext and the Laws of Media." *Postmodern Culture* 1, no. 3 (May).

——. 1989a. "In the Zones: Hypertext and the Politics of Interpretation." *Writing on the Edge* 1, no. 1: 18–27.

——. 1989b. "Hypertext and the 'Hyperreal.'" *Proceedings of Hypertext '89.* ACM, Pittsburgh.

Mouré, Erin. 1989. "Seebe and Site Glossary: Loony Tune Music." In *WSW [West South West].* Montreal: Véhicule Press.

——. 1988. *FURIOUS.* Toronto: Anansi.

Nelson, Theodor Holm. 1990. *Literary Machines.* Sausalito, Calif.: Mindful.

——. 1987a. "All for One and One for All." *Proceedings of Hypertext '87.* ACM, Chapel Hill.

——. 1987b. *Computer Lib/Dream Machines.* Redmond, Wash.: Tempus Books.

Neuwirth, Christine M., and David S. Kaufer. 1989. "The Role of External Representation in the Writing Process: Implications for the Design of Hypertext-Based Writing Tools." *Proceedings of Hypertext '89.* ACM, Pittsburgh.

Nielsen, Jakob. 1990. *Hypertext and Hypermedia.* New York: Academic Press.

Olson, Charles. 1974a. *Additional Prose: A Bibliography on America, Proprioception, and Other Essays.* Bolinas, Calif.: Four Seasons Foundation.

——. 1974b. "The Present Is Prologue." *Additional Prose: A Bibliography on America, Proprioception, and Other Essays.* Bolinas, Calif.: Four Seasons Foundation.

——. 1960. "Letter 3." *The Maximus Poems.* New York: Jargon Corinth.

——. 1947. *Call Me Ishmael.* San Francisco: City Lights.

Ong, Walter. 1982. *Orality and Literacy: The Technologizing of the Word.* New York: Methuen.

Ortega y Gasset, José. 1956. *The Dehumanization of Art.* New York: Anchor Books.

Page, Tim, ed. 1989. *The Glenn Gould Reader.* New York: Alfred A. Knopf.

Parunak, H. Van Dyke. 1991. "Ordering the Information Graph." In *McGraw-Hill Handbook of Hypertext and Hypermedia,* edited by Emily Berk and Joseph Devlin. New York: McGraw-Hill.

Pavič, Milorad. 1988. *Dictionary of the Khazars.* New York: Alfred A. Knopf.

Petry, Martha. 1993. "The Permeation of What Is In [Computer Classrooms]." Paper presented at roundtable on Computers, Writing, and the Body. CCCC, San Diego.

Plato. 1930. *The Republic.* Translated by Paul Shorey. Cambridge, Mass.: Harvard University Press.

Porush, David. 1992. "Frothing the Synaptic Bath." In *Storming the Reality Studio,* edited by Larry McCaffery. Durham, N.C.: Duke University Press.

——. 1989. "Cybernetic Fiction and Postmodern Science," *New Literary History* 20, no. 2 (Winter): 373–91.

Pound, Ezra. 1954. *Literary Essays of Ezra Pound.* New York: New Directions.

Ratcliffe, Mitch, and Dartanyan L. Brown. 1992. "Apple Wireless Devices Heard 'round the World?" *Macweek* 6, no. 5 (February): 22.

Rushdie, Salman. 1989. *Satanic Verses.* New York: Viking.

Samuelson, Pamela, and Robert Glushko. 1991. "Intellectual Property Rights for Digital Library and Hypertext Publishing Systems." *Proceedings of Hypertext '91.* ACM, San Antonio.

Sauer, Carl O. 1963. "The Education of a Geographer." In *Land and Life: A Selection from the Writings of Carl Ortwin Sauer,* edited by John Leighly. Berkeley: University of California Press.

Schank, Roger. 1982. *Dynamic Memory: A Theory of Reminding and Learning in Computers and People.* New York: Cambridge University Press.

Schank, Roger, and Robert Abelson. 1977. *Scripts, Plans, Goals, and Understanding: An Inquiry into Human Knowledge Structures.* Hillsdale, N.J.: Lawrence Erlbaum Associates.

Selfe, Cynthia L. 1989. "An Open Letter to Computer Colleagues: Notes from the Margin." Paper presented at Computers and Writing Conference, Minneapolis.

——. 1993. "Modalities of Interactivity and Virtuality." Artist's statement for a paper presented at Museum of Modern Art, Technology in the 90s series, 26 April.

Shaw, Jeffrey. 1991. "The Virtual Museum." Interactive installation.

——. 1988–91. "The Legible City." Interactive installation.

Smith, Catherine F. 1991. "Reconceiving Hypertext." In *Evolving Perspectives on Computers and Composition Studies: Questions for the 1990s,* edited by Gail E. Hawisher and Cynthia L. Selfe. Urbana, Ill., and Houghton, Mich.: NCTE Press and Computers and Writing.

Smith, Clive. 1994. "Bandwidth: Reply to Network Artists and Others Concerned (NAWOC) Initiative," electronic mail, 94-03-31, 09:02:08 est.

Stein, Gertrude. 1909. *3 Lives.* New York: Vintage Books.

Sullivan, Patricia. 1991. "Taking Control of the Page: Electronic Writing and Word Publishing." In *Evolving Perspectives on Computers and Composition Studies: Questions for the 1990s,* edited by Gail E. Hawisher and Cynthia L. Selfe. Urbana, Ill., and Houghton, Mich.: NCTE Press and Computers and Writing.

Sussman, Andrew. 1988. MS. Cornell University.

Taylor, Paul. 1990. "Hypertext, Heteroglossia, Chaos." Paper presented at Computers and Writing Conference, Austin, Tex.

Tenhaaf, Nell, and Catherine Richards. 1991. *Virtual Seminar on the Bioapparatus.* Banff: The Banff Centre for the Arts.

Thurber, D. 1990. "Sony to Make Electronic Books: 'Data Discman' Player Will Use 3-Inch CDs." *Washington Post,* 16 May, D9, D13.

Tolstoy, Leo. 1889. *The Kreutzer Sonata.* Moscow.

Trigg, Randall, L. Suchman, and Frank Halasz. 1986. "Supporting Collaboration in Notecards." *Proceedings of CSCW '86,* Austin, Tex.

Tulving, Endel. 1985. "How Many Memory Systems Are There?" *American Psychologist* 40, No. 4 (April): 385–98.

Various authors. The Dryden Statement, Broadsheet and electronic minifesto, TINAC Report, Dryden, New York and elsewhere 1988 and after.

Weinbren, Grahame. 1993a. "The Interactive Cinema." MS, 21 May.

——. 1993b. "Pointing at an Interactive Cinema." *Camerawork: A Journal of Photographic Arts* 20, no. 1 (Spring–Summer).

——. 1993c. *Sonata.* Interactive cinema installation.

Weinbren, Grahame, and Roberta Friedman. 1986. *The Erl King.* Interactive cinema installation.

Weizenbaum, Joseph. 1966. "ELIZA—A Computer Program for the Study of Natural Language Communication between Man and Machine." *Communications of the Association for Computing Machinery* 9, no. 1 (1991).

Wills, Gary. 1990. "Late Bloomer." *TIME* 135, no. 17, 23 April.

Wright, Patricia. 1991. "Cognitive Overheads and Prostheses: Some Issues in Evaluating Hypertexts." *Proceedings of Hypertext '91,* ACM, San Antonio.

Zukav, Gary. 1979. *The Dancing Wu Li Masters.* New York: William Morrow.